→INTRODUCING

STEPHEN HAWKING

J.P. MCEVOY & OSCAR ZARATE

Previously published in the
UK in 1999 by Icon Books Ltd.

This edition published in the UK
in 2009 by Icon Books Ltd.,
Omnibus Business Centre,
39–41 North Road, London N7 9DP
email: info@iconbooks.co.uk
www.introducingbooks.com

Sold in the UK, Europe, South Africa
and Asia by Faber and Faber Ltd.,
Bloomsbury House, 74–77 Great
Russell Street, London WC1B 3DA
or their agents

Distributed in the UK, Europe, South
Africa and Asia by TBS Ltd.,
TBS Distribution Centre,
Colchester Road, Frating Green,
Colchester CO7 7DW

This edition published in Australia
in 2009 by Allen & Unwin Pty. Ltd.,
PO Box 8500, 83 Alexander Street,
Crows Nest, NSW 2065

Previously published in the UK in 1995
and Australia in 1996 under the title
Stephen Hawking for Beginners and in
1999 under the current title

Reprinted 1996, 1997, 1998, 2000, 2002

This edition published in the USA
in 2009 by Totem Books
Inquiries to: Icon Books Ltd.,
Omnibus Business Centre,
39–41 North Road,
London N7 9DP, UK

Distributed to the trade in the USA by
National Book Network Inc.,
4501 Forbes Boulevard, Suite 200,
Lanham, Maryland 20706

Distributed in Canada by
Penguin Books Canada,
90 Eglinton Avenue East, Suite 700,
Toronto, Ontario M4P 2Y3

ISBN: 978-184831-094-0

Originating editor: Richard Appignanesi

Printed by Gutenberg Press, Malta

The Luckiest Man in the Universe

On 19 October 1994, the author of this book interviewed Stephen Hawking. He began with a question that might seem daring, if not impertinent. Did Hawking consider himself lucky?

I agree I have been very fortunate in everything except getting motor neurone disease. And even the disease has not been such a blow. With a lot of help, I have managed to get round the effects. I have the satisfaction in having succeeded in spite of it.
I'm really much happier than I was before it began. I can't say it has been a benefit, but I have been lucky that it has not been the disadvantage it could have been.

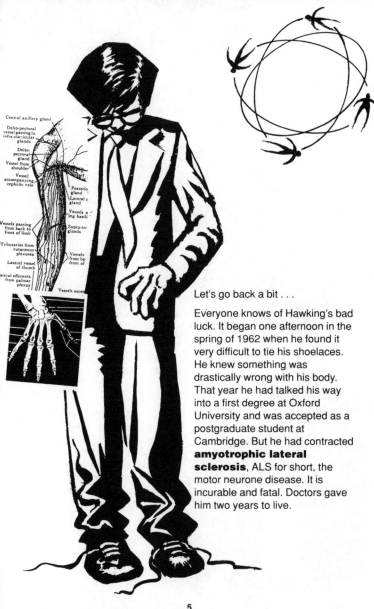

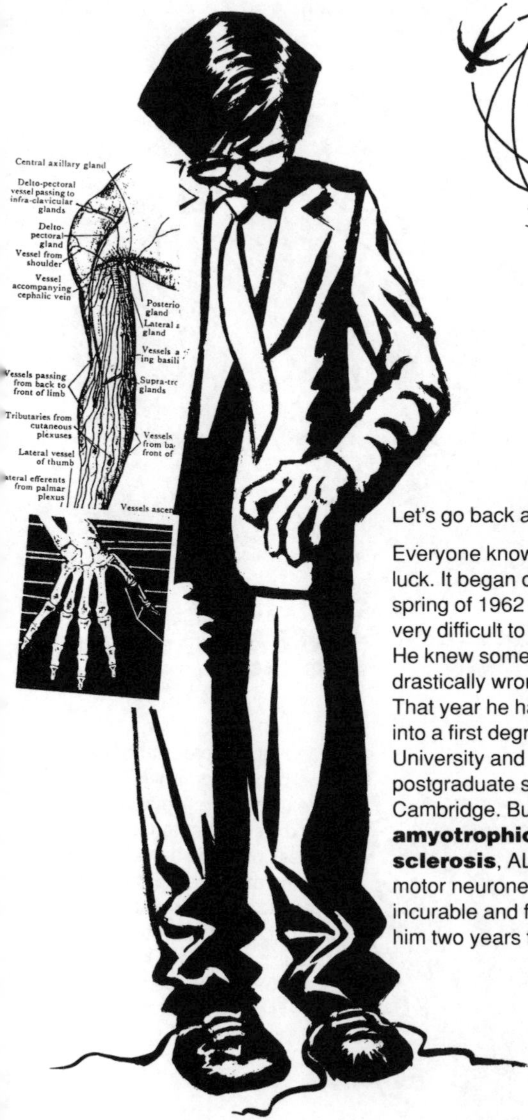

Let's go back a bit . . .

Everyone knows of Hawking's bad luck. It began one afternoon in the spring of 1962 when he found it very difficult to tie his shoelaces. He knew something was drastically wrong with his body. That year he had talked his way into a first degree at Oxford University and was accepted as a postgraduate student at Cambridge. But he had contracted **amyotrophic lateral sclerosis**, ALS for short, the motor neurone disease. It is incurable and fatal. Doctors gave him two years to live.

As the tabloid press and the paperback biographies would have us believe, Hawking spent the next several months in deep depression in his university digs, drinking and listening to Wagner. To add to his bitterness, he was told that he would not have the famous cosmologist Fred Hoyle (b. 1915) as his research adviser, the reason he chose Cambridge in the first place.

But immediately his luck began to change. A young woman, Jane Wilde, he met on New Year's Eve 1962 had taken a genuine interest in him, and the Cambridge Physics Department had assigned him to Dennis Sciama (b. 1926), one of the best-informed and most inspiring research advisers in the world of relativistic cosmology.

Once it is accepted that Stephen William Hawking's physical capabilities were severely limited by the tragic disease of ALS, a whole series of fortunate events seemed to have taken place in the early 1960s which enabled him to fulfil his destiny as one of the leading cosmologists of modern times.

First of all, for the profession he had chosen – theoretical physics – the only facility he **absolutely** needed was his brain, which was completely unaffected by his illness. He had met a helpful partner in Jane Wilde and been presented with a sympathetic thesis adviser, Sciama.

Soon he would meet Roger Penrose (b. 1931), a brilliant mathematician working on black holes, who would teach him radically new analytical tools in physics. Penrose would help him solve a research problem that would not only save his doctoral dissertation but also bring him directly into mainstream theoretical physics.

The help of these three people at such a critical time in Hawking's life is perhaps more than anyone can hope for.

1916. № 7.

ANNALEN DER PHYSIK.

VIERTE FOLGE. BAND 49.

1. *Die Grundlage der allgemeinen Relativitätstheorie;*
von A. Einstein.

He had another appointment with destiny at about the same time. A theory which had been developed almost fifty years earlier – Einstein's general theory of relativity – was only just being widely applied to practical problems in cosmology. It seems that predictions based on this theory were so bizarre that it had taken decades for it to be accepted. Now in the early 1960s, a golden age of research in cosmology based on general relativity was about to begin. Fate had waited for Stephen Hawking. The secretly ambitious – though by then slightly crippled – theoretical physicist was ready. He didn't know how long he had to live . . . but he was certainly in the right place at the right time.

Stephen Hawking is called a **relativistic cosmologist**. This means he studies the Universe as a whole (cosmologist) and uses mainly the theory of relativity (relativistic).

As Hawking has spent his entire career as a theoretical physicist – from the early 1960s to the mid 1990s – working with Einstein's general relativity, it might be a good idea to know what it's about.

The General Theory of Relativity

Berlin, November 1915. Albert Einstein (1879-1955) had just completed his theory of general relativity, a mathematical structure in which curved space and warped time are used to describe gravitation. All modern cosmology began two years later, when Einstein published a second paper called **Cosmological Considerations** in which he applied his new theory to the entire Universe.

General relativity is difficult to master, but the relatively few people who understand the theory agree it is an elegant, even beautiful theory of gravitation.

Describing a set of equations as beautiful doesn't help much in understanding how Einstein's theory differs from that of Isaac Newton (1642-1727). But an example of how each of the two theories describes gravity in the same physical situation might do the trick.

Why does a cosmologist have to study gravitation?

Cosmology is the study of the whole Universe and much of the subject is based on wide sweeping hypotheses. Gravitation determines the large scale structure of the Universe or, more simply, keeps the planets, star and galaxies together. It is the most important concept for work in this field.

Until recently, the subject of cosmology was considered to be a pseudo science reserved for retired emeritus professors. But in the last three decades, more or less coinciding with Hawking's career, two major developments have changed the subject dramatically.

■ First major breakthroughs in observational astronomy reaching out to the most distant galaxies have made the Universe a laboratory to test cosmological models

■ Second Einstein s general relativity has been proven over and over again to be an accurate and reliable theory of gravitation throughout the entire Universe

Remember physics is a cumulative subject New theories are built on previous ones keeping the ideas that stand up to experimental test and discarding those that don t Our final goal is to understand the contributions of Stephen Hawking who has taken Einstein s gravitation theory to its ultimate limit

It is important to understand the notion of *partial theories* For example. Newton s Law of Gravitation is very accurate only when gravity is weak and must be replaced by Einstein s general relativity in strong gravitational fields Similarly. relativity must be replaced by quantum mechanics when examining interactions on a microscopic scale. such as the big bang singularity or at the edge and centre of a black hole Hawking is generally recognized as the theoretician with the best chance of combining general relativity and quantum mechanics to produce quantum gravity ill named by the media as **The Theory of Everything**

THE COMPLETE STORY TAKES IN NEWTON, THEN EINSTEIN, THEN HAWKING.
FIRST, NEWTON.

Newton: The Concept of Force

Newton introduced the concept of a gravitational *force* of attraction and stated that the mutual pull of attraction between two objects is proportional to the **mass** of each object (i.e. the amount of matter the object contains) and inversely proportional to the square of distance between them.

gravitational constant — masses of two objects

$$F = G \frac{M_1 \quad M_2}{R^2}$$

distance between masses

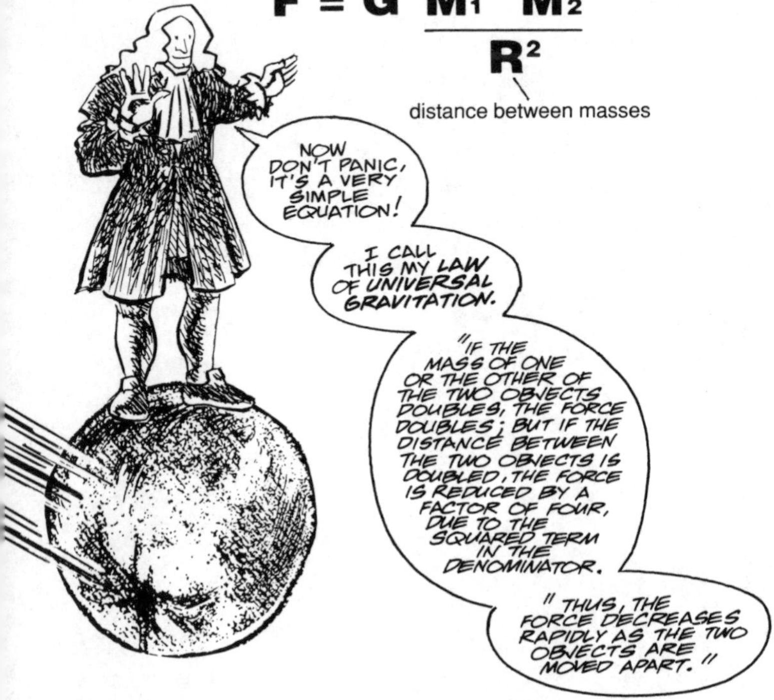

NOW DON'T PANIC, IT'S A VERY SIMPLE EQUATION!
I CALL THIS MY LAW OF UNIVERSAL GRAVITATION.
"IF THE MASS OF ONE OR THE OTHER OF THE TWO OBJECTS DOUBLES, THE FORCE DOUBLES; BUT IF THE DISTANCE BETWEEN THE TWO OBJECTS IS DOUBLED, THE FORCE IS REDUCED BY A FACTOR OF FOUR, DUE TO THE SQUARED TERM IN THE DENOMINATOR.
"THUS, THE FORCE DECREASES RAPIDLY AS THE TWO OBJECTS ARE MOVED APART."

Gravitation is the weakest force in nature as seen by the magnitude of the gravitational constant G in practical units:

$G = 6.67 \times 10^{-11}$ Newtons-metres^2 / kilograms^2

A Newton is a scientific unit of force, equal to about a quarter of a pound.

Four Kinds of Force in the Universe

The Electromagnetic Force: keeps atoms together and is the basis for all chemical reactions.

The Strong Nuclear Force: binds the neutrons and protons together in the nucleus. This force is important in nuclear reactions like fission and fusion.

The Weak Nuclear Force: determines radioactive decay, i.e. the spontaneous emission of alpha and beta particles from inside the nucleus.

The Gravitational Force: responsible for large-scale structure of the Universe, the formation of galaxies, stars, and planets.

The four known forces separate and become individually distinct during the earliest moments of the Universe.

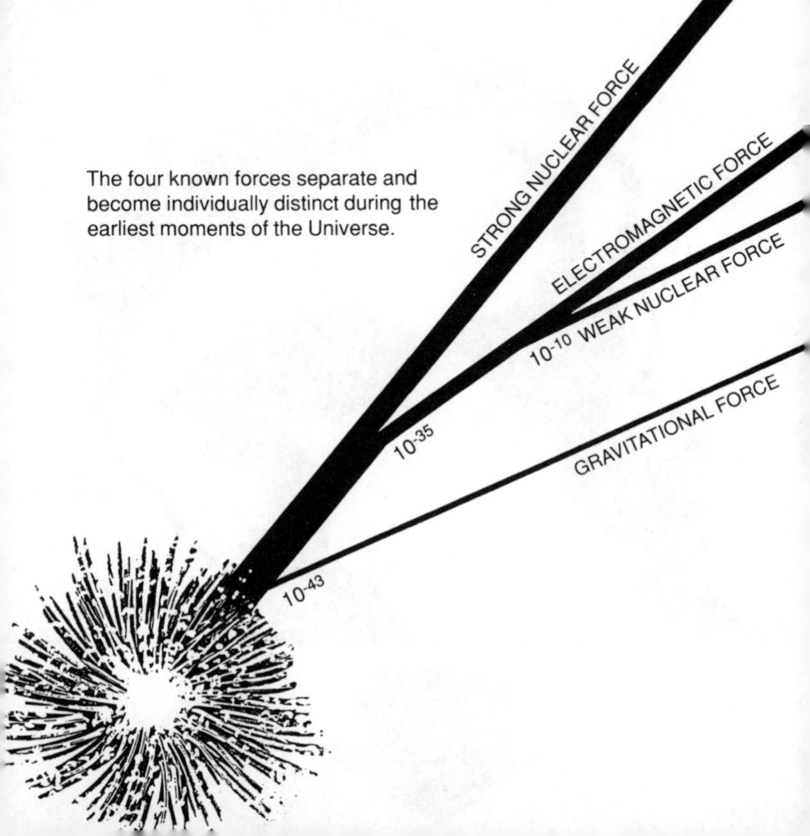

When two Sumo wrestlers (mass about 135 kilograms) get close to each other in the ring (say a metre apart), the force pushing them **towards each other** is minuscule . . . about 10,000 times less than the pull necessary to pick up one square of toilet tissue! To convert the answer to pounds multiply Newtons by 0.225.

$$\mathbf{Fg = \frac{(6.67 \times 10^{-11})(135)(135)}{(1\ metre)^2} = 0.000012\ Newtons\ (0.0000027\ lb)}$$

But the force pulling each of them **towards the floor** is much larger. That's because the other object attracting each **downwards** is the Earth, whose mass (5.98 x 10^{24} kilograms) must be put in the numerator of Newton's equation. The Earth's radius (6.37 x 10^{6} metres) goes in the denominator. Try the calculation yourself with an electronic calculator and don't forget the conversion factor to get your answer in pounds.

Fg = 298 lb (weight of Sumo)

The Principia: Describing Newton's Universe

Newton was chiefly concerned with the force of gravity between the Sun and planets, i.e., the solar system. The immediate impetus for the publication of his theory of gravitation, the **Principia**, arose from a discussion at the Royal Society in 1684 between the astronomer Edmond Halley (1656-1742), the architect Sir Christopher Wren (1632-1723) and Newton's arch rival Robert Hooke (1635-1703).

WITHOUT HESITATION, NEWTON THE RECLUSIVE GENIUS REPLIED TO HALLEY'S QUESTION OF ELLIPTICAL ORBITS.
AN INVERSE SQUARE LAW DOES INDEED PRODUCE ELLIPTICAL ORBITS, A RESULT WHICH I'VE DEMONSTRATED MATHEMATICALLY.
THIS IS VERY EXCITING! CAN YOU SHOW IT TO ME?

EVERYONE KNEW THAT JOHANNES KEPLER (1571-1630) HAD SHOWN THAT PLANETARY ORBITS WERE ELLIPTICAL, BUT A MATHEMATICAL PROOF OF IT WAS SOMETHING ELSE AGAIN.
EH, WELL... I CAN'T FIND THE CALCULATION.

Halley returned to London frustrated, but 3 months later he received a 9-page paper in Latin, **De Motu Corporum** or **On the Motion of Bodies in Orbit**, in which Newton described the elliptical paths of the planets in terms of his Law of Gravitation and his Laws of Motion. This was the precursor of his world-famous **Principia** (1687) which presented a complete mathematical description of his ideas.

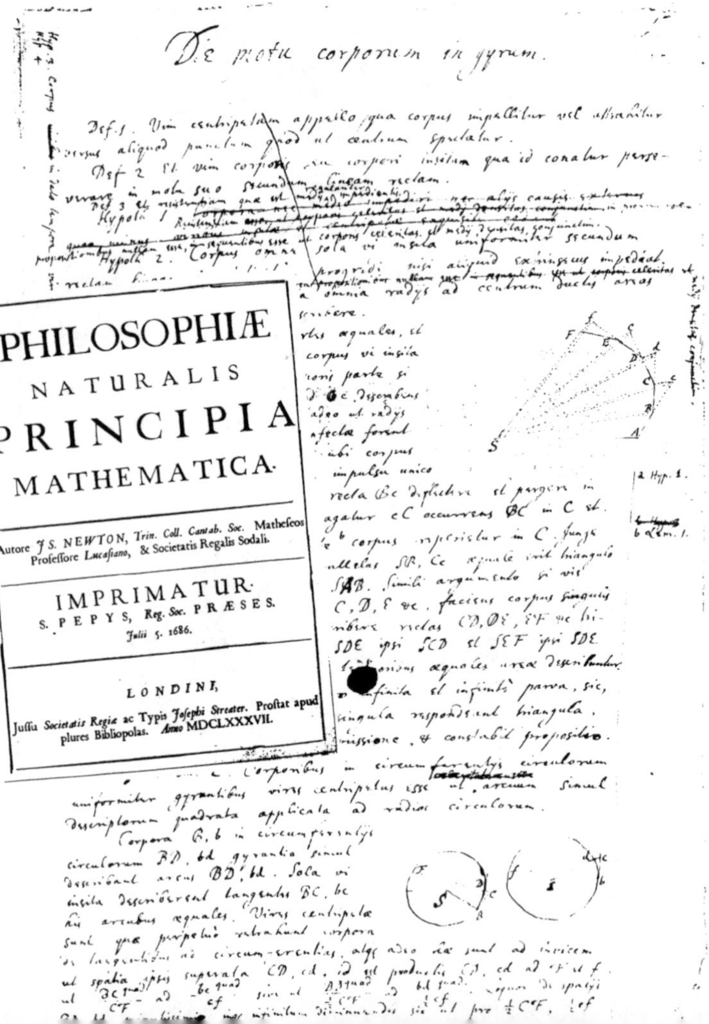

Newton and Hawking

The media often compares Stephen Hawking with other famous physicists like Newton and Einstein, much to the alienation of scientists and, in particular, historians of science. No single individual will ever dominate his era as Newton did, whereas Hawking is one of a small group of élite scientists at the cutting-edge of today's cosmology.

Yet, some of these comparisons are very interesting.

Newton spent his entire scientific career at Cambridge with his residence and laboratories at Trinity College. Hawking has been at Cambridge since his postgraduate student days in 1962, except for a few sabbatical years abroad.

They have both attempted to explain the observable physical Universe using theories of gravity: Newton using his own theory, and Hawking using mainly Einstein's general relativity.

Both have held the same distinguished position at Cambridge, the Lucasian Chair of Mathematics.

The wide range of applications for the gravitation law which Newton presented in the **Principia** is quite extraordinary. The theory was an immediate success and found to be applicable to all motion in the solar system, including the Moon and comets as well as the planets. It was so accurate that it was used to discover the planet Neptune, which could not even be seen with the telescopes available at the time.

Except for one small problem. The orbit of Mercury wasn't quite right. But as Mercury is so close to the Sun and very difficult to view, the discrepancy was thought to be due to observational errors and excused by everyone in the 17th and 18th century. The orbits of Jupiter, Mars and Saturn were spot on. No one was worried.

BUT I'M WORRIED!
QUIET! YOU DON'T EXIST YET!

Many may find it surprising to learn that putting a man on the Moon, some half-century after Einstein, did not require any modification of Newton's theory. NASA engineers were using the **Principia** when they programmed their rockets at Cape Kennedy in 1969.

But the difference is negligible, unless measurements are being made very close to a massive gravitational object. For orbits around the Sun and the planets, in fact throughout most of the entire solar system, Einstein's relativistic effects can be ignored and Newton's theory is fine.

The Concept of Mass

Consider the miracle method for losing weight: a trip to the Moon! When an object is transported in a spaceship to the Moon, its weight decreases by about a factor of 6! This weight loss can be demonstrated very simply, using Newton's gravitation formula to compare the force of gravity of a body at the surface of the Earth (i.e. its weight) with that on the surface of the Moon. Just plug the numbers into the equation and see the dramatic weight loss. But watch how you use mass.

The mass of the astronaut is about 60 kilograms (determined by a scale balance and standard masses); the **mass of the Earth** is 5.98×10^{24} kg and the **radius of the Earth** is 6.37×10^{6} metres. If we use these values in Newton's equation, we find her weight to be (with 1 Newton = 0.225 lb):

Weight = F_g = 590 Newtons = 132 lb

Now what will she weigh on the Moon? Use the same method but this time use the **mass of the Moon** = 7.34×10^{22} kilograms and the **radius of the Moon** = 1.74×10^{6} metres

Weight = 97 Newtons = 21.8 lb.

Even the Sumo will only weigh 50 lb.

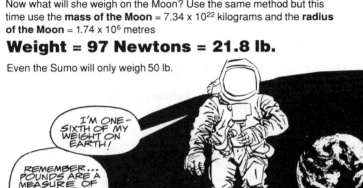

But the **mass** of the astronaut doesn't change on the Moon. She has lost none of the **matter** that makes up her body. Thus, her physical appearance and size are unaffected by the change in the gravitational field.

Mass is a tricky concept. No doubt about it. It is not only difficult to understand, but, until Einstein, it was also horribly ambiguous. Think of that property of a body that causes it to be attracted by another body, as in Newton's Law of Gravitation.

(gravitational mass)

$$F(\text{force}) = \frac{G\, m_1 m_2}{R^2}$$

Then think of the property of a body which gives it resistance to a change in speed, as in Newton's second Law of Motion;

$$\mathbf{F(force)} = \overset{\textbf{(inertial mass)}}{\mathbf{m}} \times \mathbf{a\ (acceleration)}$$

$$\textbf{or . . . } \mathbf{a} = \frac{\mathbf{F\ (force)}}{\mathbf{m\ (mass)}}$$

Clearly a larger inert mass will result in a smaller acceleration for a given force. Is there any difference between these two quantities, **gravitational** mass and **inertial** mass? Newton has confused us.

NEWTON'S A GIANT OF PHYSICAL THEORY. I'M PLEASED TO VIEW THE UNIVERSE FROM ATOP HIS SHOULDERS. BUT I'M WORRIED ABOUT *MASS...* AND A FEW OTHER THINGS!

Albert Einstein, the Saviour of Classical Physics

It was left to a single man to clear up the leftover inconsistencies of classical physics. Albert Einstein. The great Victorian physicists had decided that only trivial problems remained. Yet, Einstein proceeded to turn Newtonian physics upside down.

Imagine Newton's theoretical structure as a house of cards. It's true, Einstein only removed two of these cards. They just happened to be at the base of the structure.

I DISCARD NEWTON'S CONCEPT OF UNIVERSAL TIME AND ABSOLUTE SPACE.
...OOPS!

To do this, it was necessary to postulate that nothing can travel faster than the **speed of light**, which Einstein said was always observed to be the same. This work he called the **Special Theory of Relativity**.

Einstein's first papers were about electrodynamics and concerned with light signals and moving clocks. But he soon began worrying about gravitation and was troubled by its bewildering property of **action at distance**.

According to Newton, if the Sun disappeared in an instant, so would its gravitational field at the Earth, millions of miles away. Yet light from the Sun, with its finite speed, would continue to travel towards the Earth for another eight minutes. This troubled Einstein. So did the concept of mass.

For Einstein, such notions were paradoxes to worry about for years and years. He already knew as a young man that the hand of God was in the **details**.

Einstein the Worrier began to consider if there might be another way to explain gravity. Maybe it is not a force at all. Since the motion of a freely falling object does not depend on the object's mass or composition (as Galileo discovered in the 15th century), gravitation might be due to certain properties of the **medium** it's falling in, that is, **space** itself.

By a series of remarkably creative and idiosyncratic steps, Einstein decided that space is not flat but **curved**, and the local curvature is produced by the presence of mass in the Universe. Consequently, bodies moving through curved space do not travel in straight lines but rather follow the path of least resistance along the contours of **curved space**. These paths are called **geodesics**.

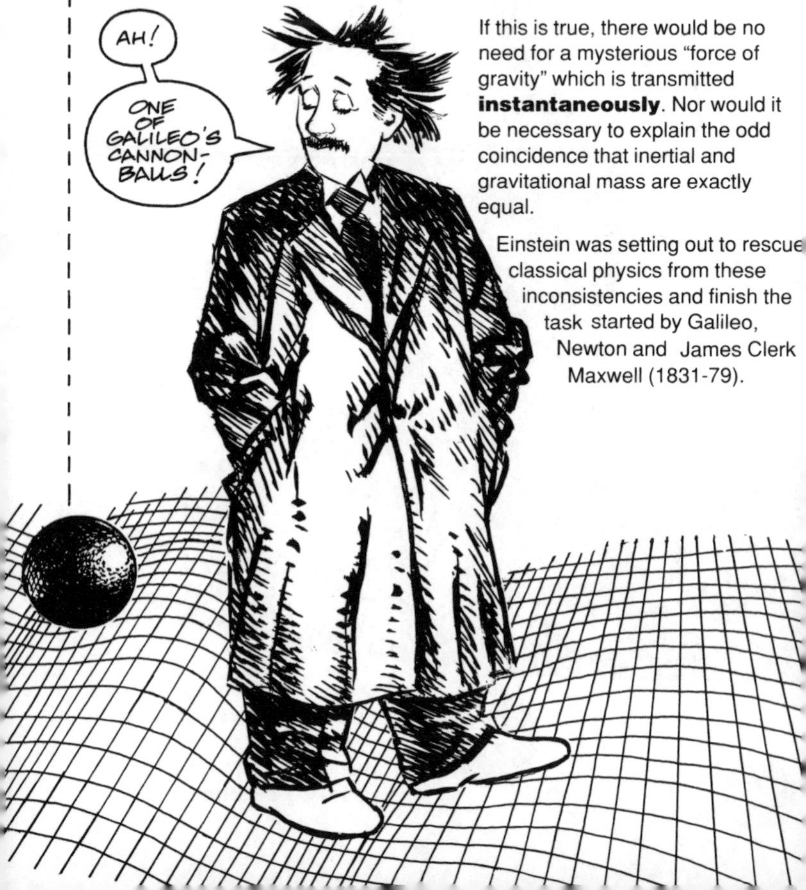
AH!
ONE OF GALILEO'S CANNON-BALLS!

If this is true, there would be no need for a mysterious "force of gravity" which is transmitted **instantaneously**. Nor would it be necessary to explain the odd coincidence that inertial and gravitational mass are exactly equal.

Einstein was setting out to rescue classical physics from these inconsistencies and finish the task started by Galileo, Newton and James Clerk Maxwell (1831-79).

Einstein and Hawking

Most great works in physics have come from those who combine miraculous physical intuition with sound mathematical skills. The former is far more important than the latter.

Einstein was not a pure mathematician and neither is Stephen Hawking. They both learned the mathematics they needed to do their physics, formulating their ideas in the most efficient way possible.

Einstein hassled his friend Marcel Grossman to learn the techniques of Riemann geometry in order to handle curved space. Hawking, anxious to probe the secrets of black holes in the early 60s, questioned Roger Penrose to exhaustion learning the new topological methods of singularity theory.

But both had a nose for the most interesting problems.

Einstein's idea of curved space had some plausibility, but it was not clear how to quantify such a new approach. So he started dreaming up more of his famous *gedanken* (thought) experiments, as he did with Special Relativity.

His sketchy qualitative ideas of curved space were to become a set of equations which gave the precise amount of curvature for a given amount of mass. This development is said to be one of the most creative examples ever of the power of pure abstract thought.

The main idea which got him started he called, **The Happiest Thought of My Life**.

Einstein's Happiest Thought

Sitting in a chair in the Patent Office at Berne (in 1907), a sudden thought occurred to me. "If a person falls freely he will not feel his own weight." I was startled and this simple thought made a deep impression on me. It impelled me towards a theory of gravitation. It was the happiest thought of my life.

I realized that . . . for an observer falling freely from the roof of a house there exists – at least in his immediate surroundings – no gravitational field. If the one who is falling drops other bodies (e.g. Galileo's cannon balls), then these remain relative to him in a state of rest or of uniform motion independent of their particular chemical or physical nature. (Of course, we are ignoring the effect of air resistance.)

The observer therefore has the right to interpret his state as at rest or in uniform motion . . .

He continued . . .

Because of this idea, the uncommonly peculiar experimental law – that in the gravitational field all bodies fall with the same acceleration (this is another way of saying that gravitational mass is the same as inertial mass) – *attained at once a deep physical meaning. If there were to exist just one single object that falls in a different way than all the others, then with its help the observer could realize that he is in a gravitational field and is falling in it. However, if such an object does* ***not*** *exist – as experience has shown with great accuracy starting with Galileo in 1590 – then the observer lacks any objective means of perceiving himself as falling in a gravitational field. He has the right to consider his state as one of rest and his environment as free of gravity. Therefore, the fact that the acceleration of free fall is independent of the nature of the material involved is a powerful argument that the relativity postulate can be extended to coordinate systems which are in* **non-uniform** *motion.*

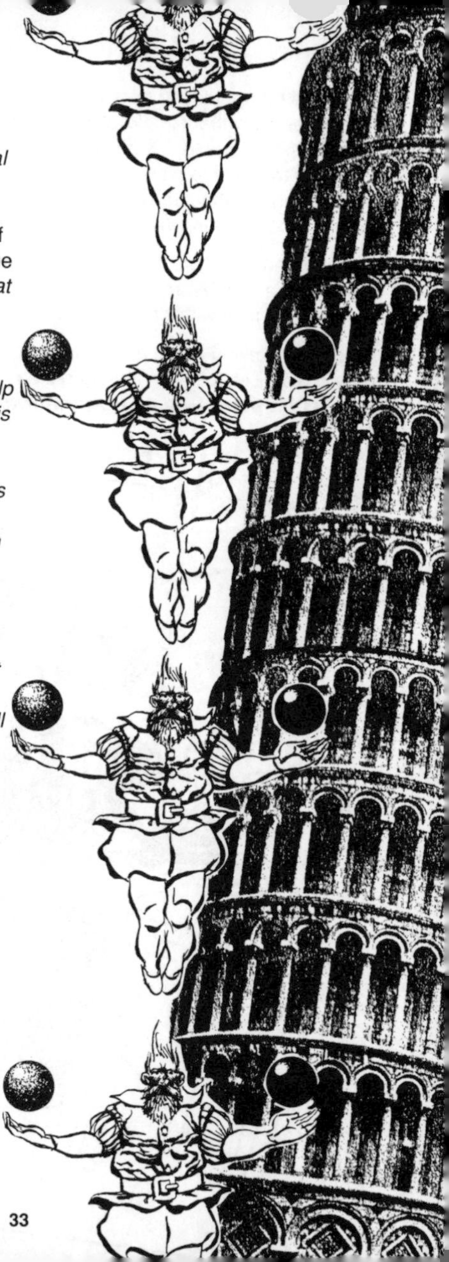

Einstein's thought that a person falling freely does not feel his own weight seems rather simple. Yet from this starting point, he squeezed every possible drop of insight, removing all the inconsistencies of Newton's theory that his intuition and the laws of physics would allow. He transformed the *simple picture of someone falling through space* into *a small laboratory in which gravity did not exist.*

He could then analyse the effect of gravity on such phenomena as the bending of a light beam or the slowing of a clock by simply replacing the gravitational field with simulated accelerated motion.

Simply by thinking about a man jumping off a roof in Berlin (or so the story goes), Einstein was able to replace gravity by acceleration and discover his **principle of equivalence**.

Einstein could now use the powerful principle of relativity – that the laws of physics should not depend on any particular reference frame – to test his new laws of space curvature. He also had the principle of equivalence (gravity equals acceleration) to get started. And he had one more useful bit of information, this one experimental.

The Perihelion of Mercury: from a Problem to a Solution

Recall that scientists in Newton's time were not worried about the small discrepancy in Mercury's elliptical orbit, even though it did not return to the same starting point in each cycle. By Einstein's time, astronomers were more than worried, they needed an explanation. The discrepancy had been carefully measured to be 43 seconds of arc per century and it would not go away. Einstein could now use the perihelion result to test his curvature law. (Perihelion, from the Greek peri meaning close to and helios, sun).

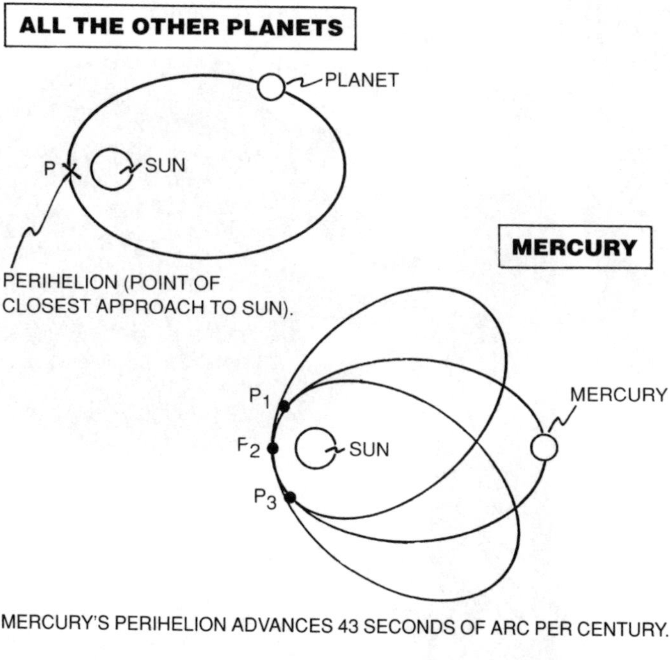

MERCURY'S PERIHELION ADVANCES 43 SECONDS OF ARC PER CENTURY.

Finding the Right Equation

Einstein used the "3 Ps" to test his equations . . .

Principle of Relativity
Perihelion of Mercury
Principle of Equivalence
He went on producing sets of equations (mentally exhausted and trying to ignore the First World War) . . .
UNTIL MY EQUATIONS FINALLY PRODUCED...
1. the correct prediction for the shift of Mercury's perihelion
2. incorporated the equivalence principle
3. and obeyed the Principle of Relativity, i.e., they had the same form when expressed in each and every reference frame he could imagine.

These latest equations also predicted a deflection of 1.7 seconds of arc for starlight passing near the edge of the Sun and incorporated his earlier prediction of gravitational time dilation, the warping of time.

Einstein presented this final form of his general relativity law of curved space and warped time to the Prussian Academy on 25 November 1915.

Then he sat down and wrote a letter to a close friend, the Dutch physicist Paul Ehrenfest.
"I WAS BESIDE MYSELF WITH ECSTASY FOR DAYS.
"IMAGINE MY JOY THAT THE NEW LAW OF CURVATURE OBEYS THE PRINCIPLE OF RELATIVITY AND PRE-DICTS THE CORRECT PERIHELION MOTION OF MERCURY.
"... THE YEARS OF SEARCHING IN THE DARK FOR A TRUTH THAT ONE FEELS BUT CANNOT EXPRESS – THE INTENSE DESIRE AND THE ALTERATIONS OF CONFI-DENCE AND MISGIVING UNTIL ONE BREAKS THROUGH TO CLARITY AND UNDERSTANDING – ARE KNOWN ONLY TO HIM WHO HAS EXPERIENCED THEM HIMSELF!"
HUPFELD

The Field Equations – What do they mean?

The 36-year old professor had produced a set of mathematical equations which gave the details of the relationship between the curvature of space and the distribution of mass in the Universe. Einstein found that matter tells space *how to curve* and then space tells matter *how to move* – a new way to describe gravitation. **No forces.** A mind flip is necessary to jump between the two pictures of gravitation.

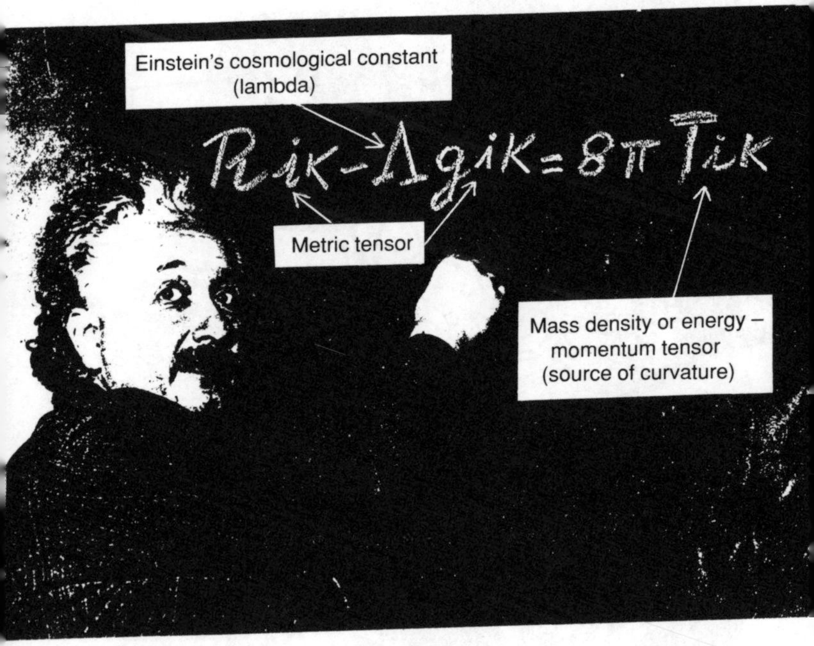

Contained within these miraculous equations is the explanation of the perihelion shift of Mercury, the degree of bending of starlight, the existence of gravitational waves, information on the singularities of space time, the description of the formation of neutron stars and black holes, even the prediction of the expansion of the Universe.

That's the good news.

The bad news is that the mathematics is extremely difficult. There are some 20 simultaneous equations with 10 unknown quantities. The equations are almost impossible to solve except in situations where symmetry or energy considerations reduce them to simpler forms.

If we ignore the cosmological constant *lambda* (which doesn't belong there anyway) and consider free space where the mass tensor is zero, the equations can be written very simply . . .

$R_{ik} = 0$
This is called the vacuum solution.

This form was made famous by a well-known photograph of Einstein lecturing on the theory in the 1920s. Looks easy!

Visualising Curved Space: the Rubber Sheet Model

Einstein's gravitation is quite unusual, compared to other field theories like electricity or magnetism, in that the description of motion (i.e. how an object moves) is already built into the field equations (how space time is curved. This can be understood with the help of a simple model – call it the rubber sheet picture.

Consider a billiard table with the slate top and felt cover replaced by a taut thin rubber sheet which is highly stretchable. If a light object (e.g. a ping-pong ball) is rolled across the sheet, it will move more or less in a straight line. This simulates *flat* space and the ping-pong ball's path corresponds to the straight line motion of *special* relativity.

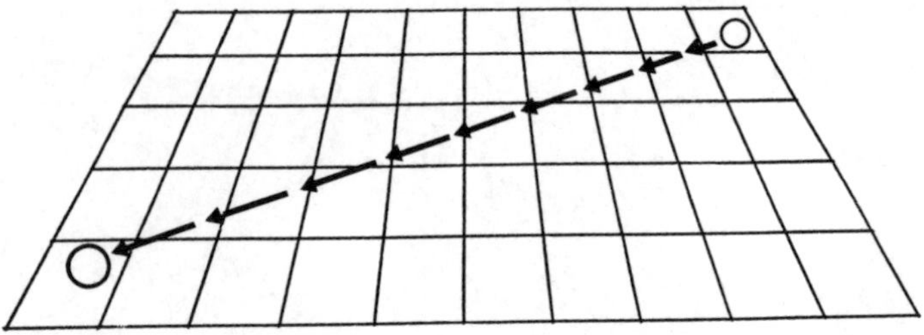

Now place a heavy billiard ball in the centre of the sheet, causing it to become curved with a depression at the centre. The model now simulates the curvature of space near a central mass as described by general relativity.

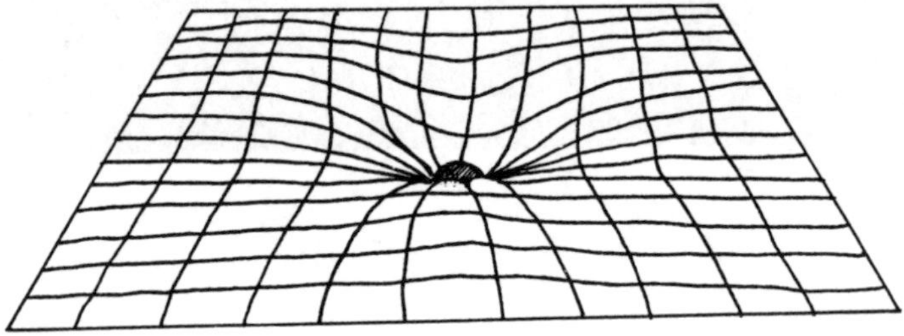

The simplest case (other than a straight line) is when the depression just captures the moving object to produce a circular orbit. Note this occurs without the need of any *centripetal force* to keep the object in orbit, as in Newton's picture.

The object would like to move in a straight line, but the space is curved, so it moves in a circle around the centre. It is simply moving along a path of least resistance in the curved space. This is general relativity's representation of how a planet is captured in an orbit around the sun.

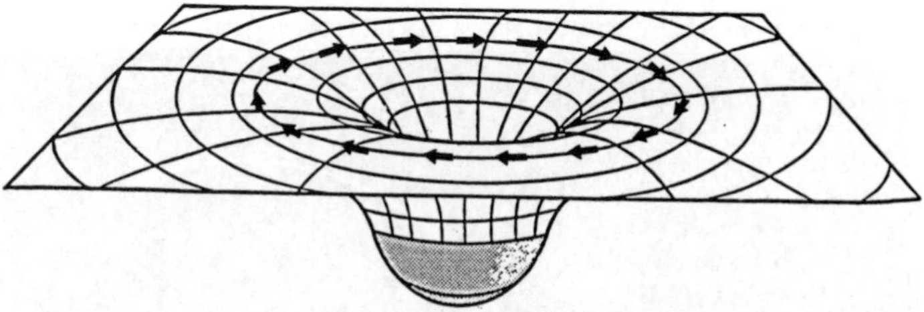

If the object is moving on a line directly towards the centre, it falls right into the depression and accelerates into the attracting centre. This is a representation of a meteorite crashing into the Sun or the Earth.

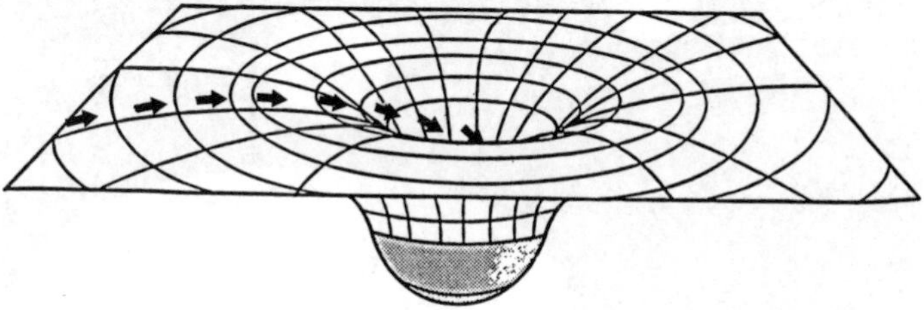

With such diagrams it is now possible to visualize the distinct and utter difference between Newton and Einstein. Einstein has replaced Newton's **gravitational force** with **curved space**.

At the time of publication, the new theory met with scepticism. Many did not wish to see the Newtonian scheme abandoned. These sceptics needed more evidence.

The Bending of Starlight: Eclipse of 29 May 1919

Four years later, the scientific world awaited the verdict of an experiment which Einstein himself had suggested in the original paper, the bending of starlight during a solar eclipse. The theory predicted that starlight passing just at the edge of the Sun would be displaced by 1.7 seconds of arc from its true position. It was the first real test of the theory.

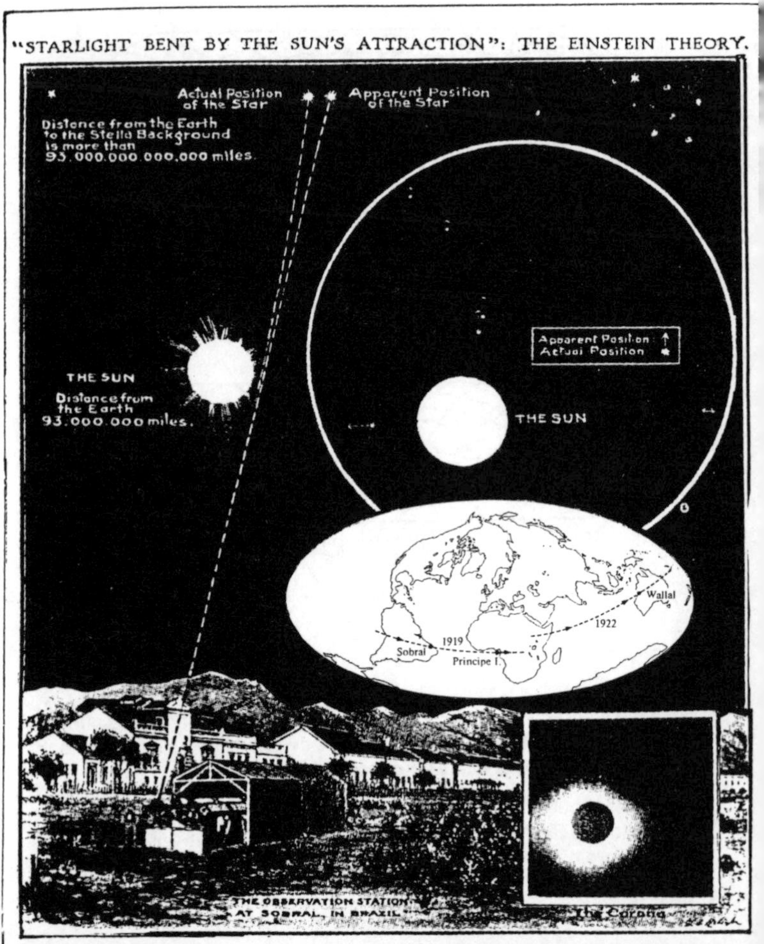

There was to be a total eclipse of the Sun on 29 May 1919, smack in the middle of a bright field of stars in the cluster Hyades. These were most unusual and optimal conditions for such an experiment.

The English astronomer **Arthur Stanley Eddington** (1882-1944) led an expedition to the island of Principe off the coast of Africa to photograph the eclipse.

Eddington found that light rays which had left the surface of stars thousands of years ago and had been bent by the curved space near the Sun only eight minutes previously, passed through the lens and exposed the photographic plates just where Einstein said they would. One of the most remarkable experiments in scientific history had been completed.

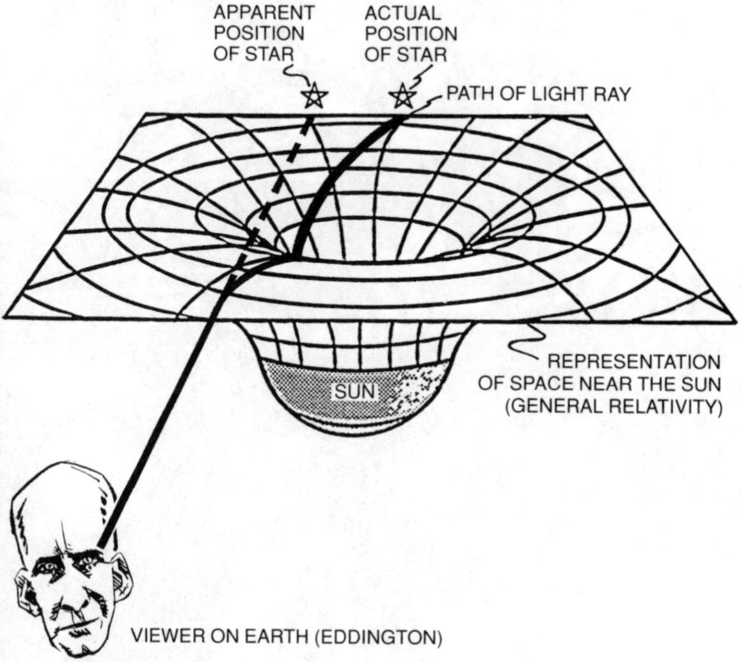

The two dimensional rubber sheet drawing of the star displacement makes it look so very simple.

The results of the eclipse expedition were presented by the Astronomer Royal at a meeting at the Royal Society on 6 November 1919, and Einstein became an international hero overnight. Headlines in the *New York Times* suggested that a new Universe had been discovered . . . and this time the newspaper hype was not exaggerated.

A world weary from war embraced the quiet and eccentric scientist, sitting in his study in Berlin with a pencil and pad, who had figured out the great plan of the Almighty for the entire Universe.

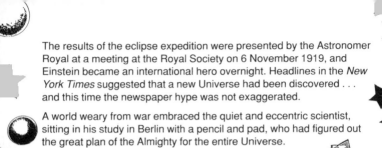

Many critics said the results were inconclusive, that the possibility of error in the star measurements was too great . . . so the scepticism continued.

Solving Einstein's Equations: Hawking's Starting Material

In the 25 years between the publication of Einstein's general relativity and the outbreak of the Second World War, several solutions to the field equations were produced which have been fundamental to Stephen Hawking's work.

IT IS A REMARKABLE FACT THAT ALL OF THESE RESULTS WERE IGNORED OR RIDICULED WHEN THEY WERE PUB-LISHED; PARTIC-ULARLY BY THE CREATOR OF THE THEORY HIMSELF, *ALBERT EINSTEIN.*
THE FIRST OF THESE SOLUTIONS APPEARED ALMOST IMMEDIATELY.

1) The Schwarzschild Geometry

In 1915, the same year as Einstein's publication, the German mathematician Karl Schwarzschild sent a paper to Einstein. Schwarzschild used elegant mathematical analysis to produce an exact solution to the equations for an arbitrary spherical body, like a star. The solution intrigued Einstein greatly because he himself had only been able to arrive at an **approximate** solution to his own equations and thought that an **exact** solution of the equations would never be found.

Schwarzschild's solution was quite an achievement because of the technical manipulation required to solve a system of ten equations connecting twenty quantities, resulting in hundreds of terms. These are not simple algebraic equations, but second order, non-linear, partial differential equations – the bane of all graduate students in physics.

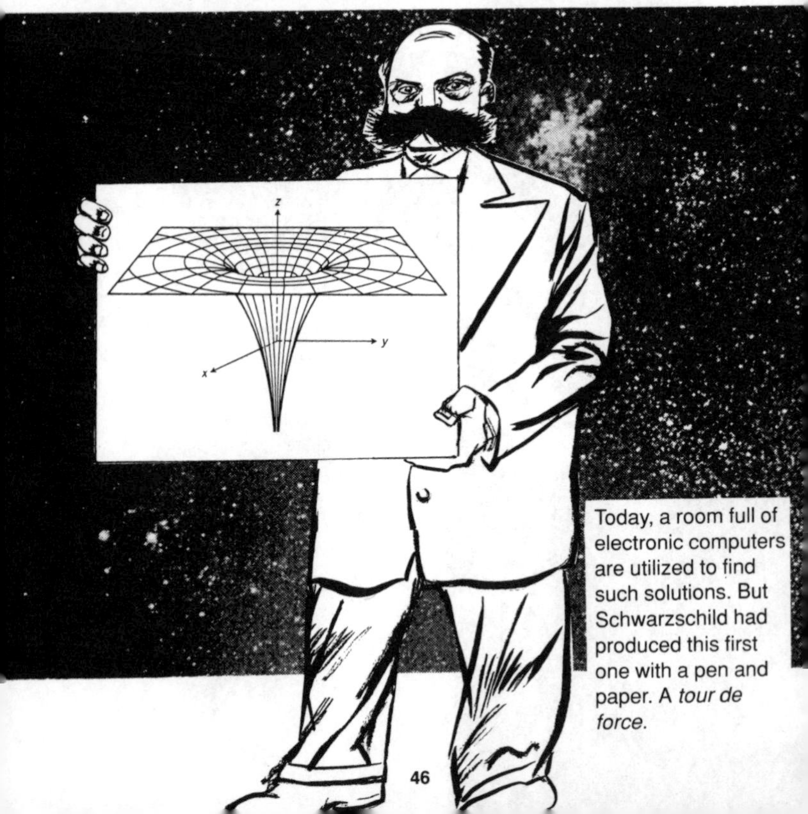

z
y
x
Today, a room full of electronic computers are utilized to find such solutions. But Schwarzschild had produced this first one with a pen and paper. A *tour de force*.

The Critical Radius

Schwarzschild's mathematics showed how the space curvature around an object of any arbitrary mass varied as a function of the distance from the centre of the object, i.e. along a **radial line.**

His results produced a very strange geometry. There seemed to exist a critical point at which the curvature was so strong that matter could not escape. This critical point is now known as the **Schwarzschild Radius** and depends only on the mass of the object. (G is the gravitational constant; c is the speed of light)

$$R = \frac{2GM}{C^2} \qquad \text{(Schwarzschild Radius)}$$

There was no immediate concern about this critical point, since the interior of stars and planets could not be investigated anyway. But there was speculation as to what might happen if a star or planet existed which satisfied this equation. The gravitational forces would be so great that the object would collapse indefinitely and **nothing** would be able to resist the self-gravity caused by the extreme space curvature. All the matter would be compressed to a singularity – a single point at the centre.

Planets as massive as the Earth would have to be compressed to absurd dimensions – to the size of a garden pea or the Sun to a diameter of about 3 kilometres. Ridiculous, they said. The calculation was a mathematical fluke. In any case, nobody wanted to think about it. Least of all, Einstein.

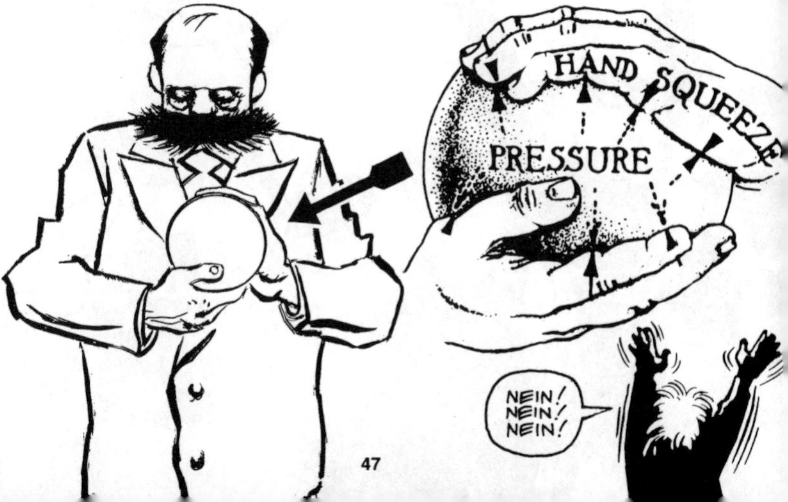

2) Friedmann: the Expanding Universe

Some years after Schwarzschild, another controversial solution to Einstein's equations appeared. In 1922, the Russian Alexander Friedmann (1888-1925) made the simplifying assumption that the Universe was **uniformly** filled with a thin soup of matter. (Modern measurements have shown this assumption of uniformity to be quite reasonable in spite of the formation of stars and galaxies.)

Friedmann found that general relativity predicted the Universe to be unstable and the slightest perturbation would cause it to expand or contract. He corrected a mistake in Einstein's 1917 paper on cosmology to reach this result. (Any wonder Einstein didn't like **this** prediction.)

Recall that Einstein had introduced an artificial term (*lambda*, the cosmological constant) into his field equations essentially to "stop the expansion". At the time, astronomers were telling him that the Universe was static, so he wanted to guarantee the theory would agree with observations. Later, he called this "cosmological constant" the biggest mistake of his life.

Friedmann dropped the *lambda* from the equations and got an **expanding universe**, which, of course, Einstein did not like. This was another solution of his own equations which he ridiculed.

Friedmann's predictions for the expansion of the Universe can be summarized by considering three different values for the mass of the Universe in terms of a ratio Ω (omega).

- **Mass density of the Universe is greater than the critical value**

In this case, the expansion rate is slow enough and the mass great enough for gravity to stop the expansion and reverse it. A Big Crunch would eventually occur with all the matter in the Universe pulled back to a single point. $\Omega > 1$ (greater than . . .)

- **Mass density of Universe is less than the critical value**

The Universe expands much more rapidly. Gravity can't stop it, but does slow the rate of expansion somewhat. $\Omega < 1$ (less than . . .)

- **Mass density of the Universe is equal to the critical value**

The Universe expands just fast enough not to collapse. The speed at which the galaxies recede from each other gradually decreases, but galaxies always move apart. $\Omega = 1$ (equal)

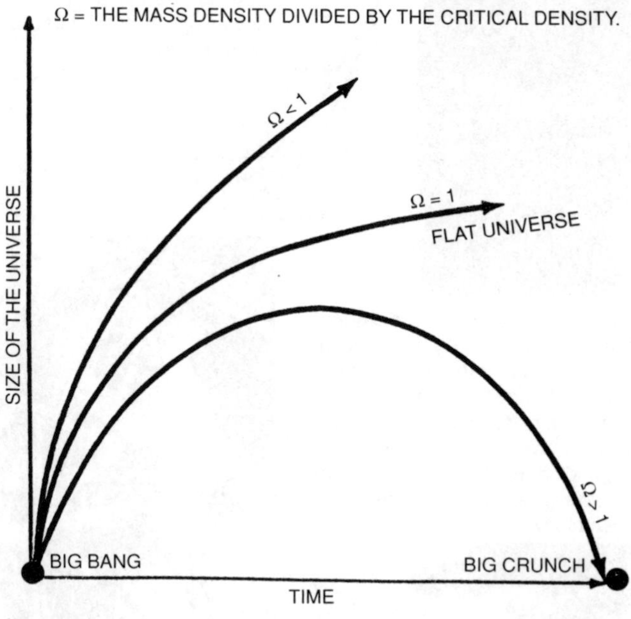

Precursor to the Big Bang: Lemaître's Primordial Aim

YEP, IF THEY'RE NOT RIGHT, I AM!

The Belgian cosmologist **Abbé Georges Lemaître** (1894-1966) was the first to use Friedmann-type solutions to formulate a model for the beginning of the Universe which he called the Primordial Atom or Cosmic Egg.

Lemaître was a visionary. Not only did he anticipate that the expanding Universe would be confirmed by looking for red shifts in the spectra of galaxies, but he even suggested that it might be possible to detect remnant radiation from the primordial atom. These two ideas dominate contemporary Big Bang cosmology in this last decade of the 20th century.

By 1929, the astronomer **Edwin Hubble** (1889-1953) had used the 100-inch Hooker telescope at the Mount Wilson Observatory in California to discover galaxies and confirm that the Universe **is** expanding. But he knew nothing of Einstein's theory or Lemaître's cosmology.

At last in 1931, Lemaître cornered Hubble and Einstein at Caltech and gave a seminar on his model Universe.

3) Oppenheimer: on Continued Gravitational Collapse, 1939

The third solution of Einstein's equations, important to modern cosmology and Stephen Hawking, was published by the American physicist J. Robert Oppenheimer (1904-1967) and one of his students, Hartland Snyder in 1939. They took up the problem of the Schwarzschild geometry in spite of the criticism by Einstein, Eddington and just about everybody else. The paper, which was published in *The Physical Review* was titled, "On Continued Gravitational Collapse".

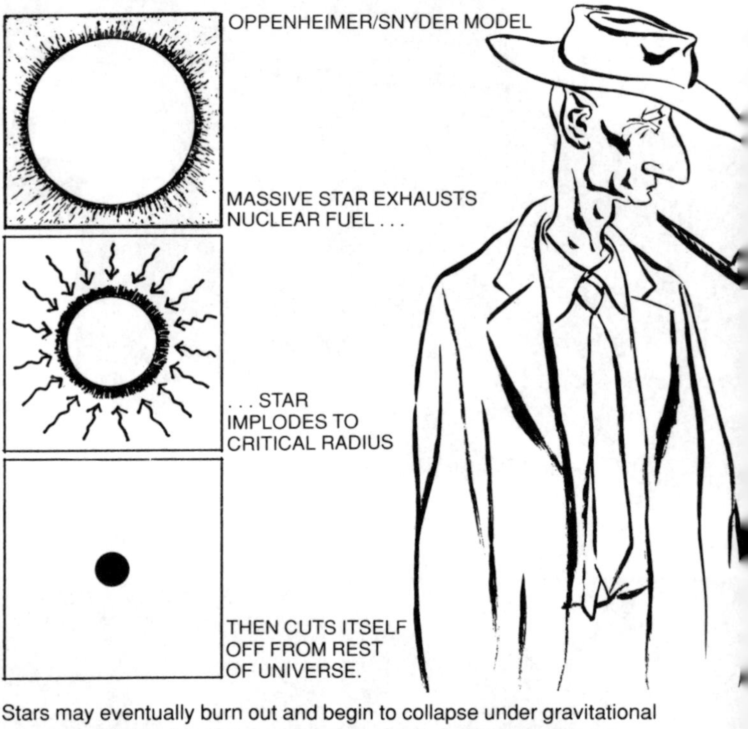
OPPENHEIMER/SNYDER MODEL
MASSIVE STAR EXHAUSTS NUCLEAR FUEL . . .
. . . STAR IMPLODES TO CRITICAL RADIUS
THEN CUTS ITSELF OFF FROM REST OF UNIVERSE.

Stars may eventually burn out and begin to collapse under gravitational contraction. In the idealized model of a spherical contracting star, a squeezing phenomenon can occur which could bring the star to the critical radius R_c. Catastrophic gravitational collapse would take place for the critically collapsed star.

■ Space curvature would be so severe that light rays emitted from the star's surface would bend into the star's interior, sealing off events from external observers.

■ Light rays at the surface would be infinitely red-shifted, i.e., the light would have no energy.

■ A one-way event horizon would form in which particles, radiation, etc. could enter the star, but nothing could be emitted.

■ A space-time singularity would ultimately form, not at the critical radius, but at the centre of the star. All the physics is continuous for an observer falling in with the collapsing star's surface.

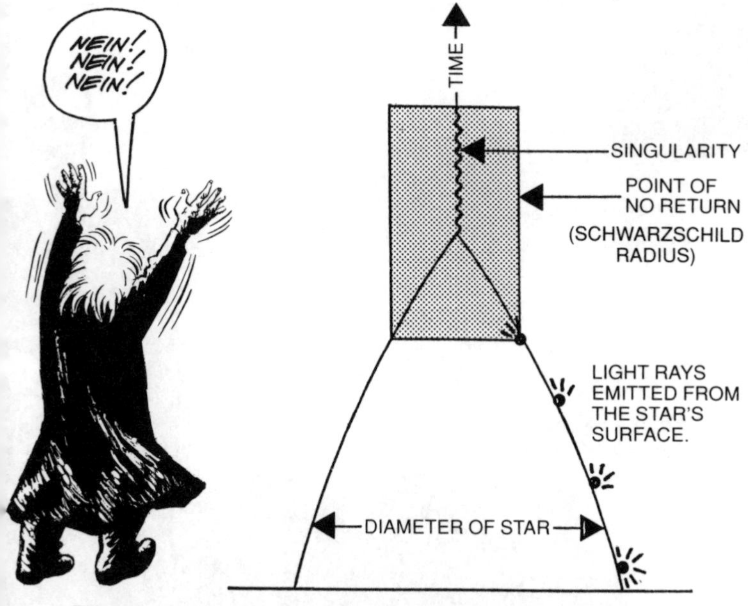

Einstein again resisted. He ridiculed the Oppenheimer result vigorously in print.

He even refused to accept that relativity could describe collapsed stars which did **not** become critical – called neutron stars – in spite of independent predictions by the eccentric Fritz Zwicky (1898-1974) at Caltech and the highly respected Lev Landau (1908-68) in Moscow.

1 September 1939

■ Publication date for the **Physical Review** issue containing article by Oppenheimer (and Snyder) describing the gravitational collapse of a star.

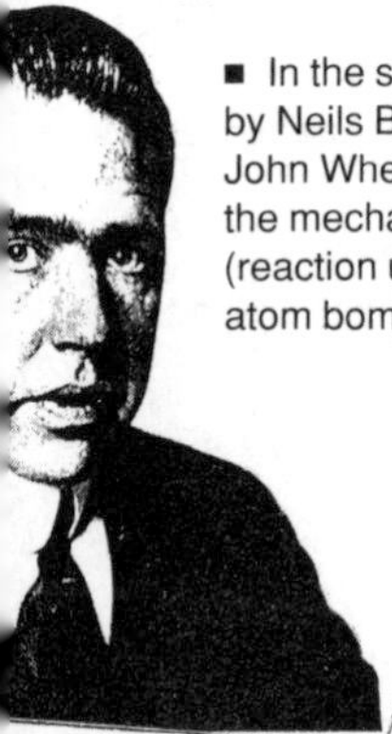

■ In the same issue, another article by Neils Bohr (1885-1962) and John Wheeler (b. 1911) explained the mechanism for nuclear fission (reaction used in the atom bomb).

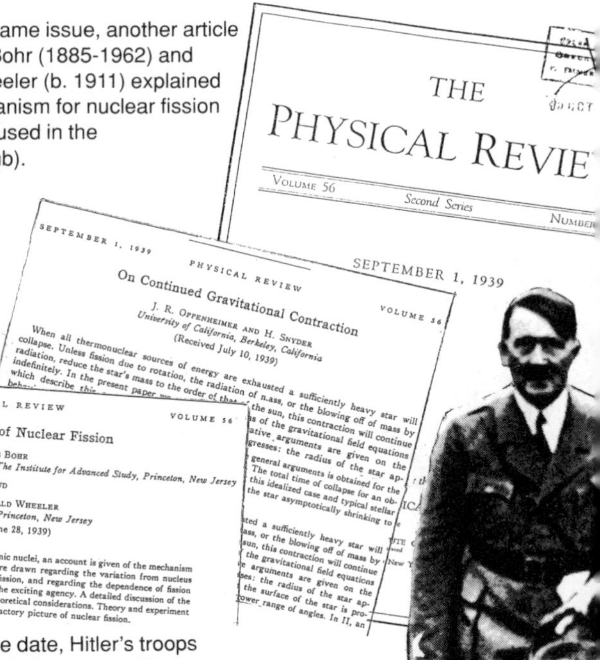

THE PHYSICAL REVIE

VOLUME 56 Second Series NUMBER

SEPTEMBER 1, 1939

SEPTEMBER 1, 1939 PHYSICAL REVIEW VOLUME 56

On Continued Gravitational Contraction

J. R. OPPENHEIMER AND H. SNYDER
University of California, Berkeley, California
(Received July 10, 1939)

When all thermonuclear sources of energy are exhausted a sufficiently heavy star will collapse. Unless fission due to rotation, the radiation of mass, or the blowing off of mass by radiation, reduce the star's mass to the order of that of the sun, this contraction will continue indefinitely. In the present paper we ... of the gravitational field equations which describe this ... arguments are given on the ... the radius of the star ap- ... general arguments is obtained for the ... The total time of collapse for an ob- ... this idealized case and typical stellar ... the star asymptotically shrinking to ...

BER 1, 1939 PHYSICAL REVIEW VOLUME 56

The Mechanism of Nuclear Fission

NIELS BOHR
ity of Copenhagen, Copenhagen, Denmark, and The Institute for Advanced Study, Princeton, New Jersey

AND

JOHN ARCHIBALD WHEELER
Princeton University, Princeton, New Jersey
(Received June 28, 1939)

On the basis of the liquid drop model of atomic nuclei, an account is given of the mechanism of nuclear fission. In particular, conclusions are drawn regarding the variation from nucleus to nucleus of the critical energy required for fission, and regarding the dependence of fission cross section for a given nucleus on energy of the exciting agency. A detailed discussion of the observations is presented on the basis of the theoretical considerations. Theory and experiment fit together in a reasonable way to give a satisfactory picture of nuclear fission.

■ On the same date, Hitler's troops invaded Poland, triggering off the Second World War.

When nuclear fission was discovered by the Germans Otto Hahn (1879-1968) and Fritz Strassman (b. 1902), physicists and politicians in Western democracies became alarmed that the Germans were developing an atom bomb to turn the entire world into a Nazi empire, a Third Reich ruling with the threat of nuclear devastation.

It is easy to see why work on cosmology was postponed. Contemplating the mysteries of the physical Universe in such severe political crisis was a luxury the free world could not afford.

In addition, the originator of the general theory had opposed all the radical cosmological predictions of his own equations as developed by Schwarzschild, Friedmann and Oppenheimer. It would be 20 years before this work was resumed and the consequences of these solutions appreciated.

1942 . . . A Turning Point in the Story

In 1942, physicists began to focus on deadly practical projects. Oppenheimer, one of the heroes of early cosmological research, left the heady intellectual climate of Berkeley for the barren flats of Los Alamos and the Manhattan Project. In December 1942, the Italian Enrico Fermi (1901-54) and his team at the University of Chicago achieved the first controlled nuclear chain reaction.

And at the beginning of that same year, on 8 January, Stephen William Hawking was born in Oxford. His mother had just moved from London to escape the nightly pounding by the German Luftwaffe.

Research on collapsing stars was abandoned for over twenty years, enough time for Stephen Hawking to grow to maturity, finish his degree at Oxford and enrol as a postgraduate student at Cambridge University.

The Death of Einstein

Albert Einstein died on 18 April 1955 in Princeton, a small college town in New Jersey, USA. His wish was to cremated so that "no one will worship at my bones". In spite of his wish, unethical doctors performed an unnecessary autopsy and made off with his brain and his eyes – an insidious invasion of privacy.

Einstein had left Europe for the USA in 1933 with his real creative work behind him. During the last 22 years of his life, he did not work on any of the important cosmological questions which came out of his general relativity theory. For years he stuck slavishly to the task of trying to unite the field equations of general relativity with Maxwell's equations for the electromagnetic field and ignored quantum mechanics

His unified field theory calculations were found by his bedside

Two other physicists who also lived in Princeton mourned the death of the great scientist. Oppenheimer, no longer affiliated with the war effort, was director of the Institute of Advanced Studies (where Einstein had an honorary position) and John Wheeler was Professor of Physics at Princeton University. Wheeler had recently finished critical years of development on the hydrogen bomb and was now returning to basic research in cosmology, with particular interest in collapsing stars.

How fitting that these two physicists should live on **opposite** sides of the same street in this small academic community. They had vastly different views of the Universe **and** of American political life which placed them on opposite sides of controversial issues, like national security and nuclear weapons. Soon they would confront each other again on the question of general relativity and gravitationally collapsing stars.

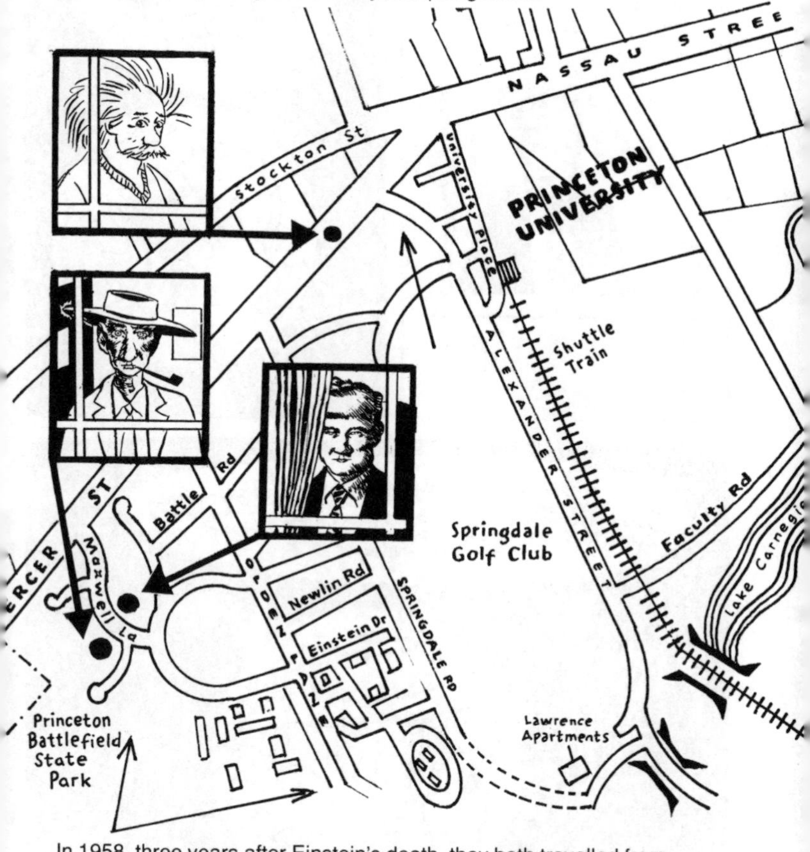

In 1958, three years after Einstein's death, they both travelled from Princeton to attend an international conference in Brussels on modern cosmology. Wheeler had been invited to give a talk reviewing the current state of research.

OF ALL THE IMPLICATIONS OF GENERAL RELATIVITY, THE QUESTION OF THE FATE OF GREAT MASSIVE STARS IS ONE OF THE MOST CHALLENGING. BUT IMPLOSION AS CALCULATED BY OPPEN-HEIMER IN 1939 DOES NOT GIVE AN ACCEPT-ABLE ANSWER.
WHY NOT? IF STARS MUCH HEAVIER THAN THE SUN REALLY OCCUR IN STELLAR EVOLUTION, THEN I BELIEVE THEIR COL-LAPSE CAN BE DE-SCRIBED IN THE FRAMEWORK OF GENERAL RELATIVITY.
WOULD NOT A SIMPLE ASSUMPTION BE THAT SUCH MASSES UNDERGO CONTINUAL GRAVITATIONAL CONTRACTION AND ULTIMATELY CUT THEMSELVES OFF FROM THE REST OF THE UNIVERSE?

REAL STAR
OPPENHEIMER + SNYDER'S IDEALIZED STAR
OPPIE, YOUR MODEL IS JUST TOO IDEALISED. REAL STARS UNDER CON-TRACTION WOULD PRODUCE NUCLEAR REACTIONS, SHOCK WAVES, HEAT RADIATION AND MASS EJECTION...
ALL SORTS OF UNPREDICTABLE PHENOMENA!

IF ONLY I'D KEPT THOSE COMPUTER SIMULATIONS ON THE EX-PLOSION OF THE HYDROGEN BOMB, WE COULD HAVE REWRITTEN THE PROGRAMS TO SIMULATE THE IM-PLOSION OF A MASSIVE STAR.
HMMM... THEY STILL HAVE BOMB DESIGN EXPER-TISE AT LOS ALAMOS AND AT LIVERMORE.
MAYBE THEY WOULD HELP WITH THE CALCULATIONS.

A few years later, Edward Teller phones Wheeler from the Livermore Radiation Labs in California.

JOHN, DR COLGATE AND HIS TEAM HAVE COMPLETED THE SIMULATION OF COL-LAPSING STARS YOU REQUESTED. IT TURNS OUT THAT WHEN A STAR IS SMALL, A SUPERNOVA IS TRIGGERED AND A NEUTRON STAR MOST LIKELY FORMS.
BUT FOR STELLAR MASSES GREATER THAN TWICE THE SUN'S MASS, THE COMPUTER SIMULATIONS SHOW THAT IMPLOSIONS PRODUCE CONTINUED GRAVITATIONAL COLLAPSE, AS PREDICTED BY THE RELATIVITY THEORY.
GREAT, JUST LIKE OPPIE SAID. WAIT TILL HE HEARS HE WAS RIGHT AFTER ALL.

Five years later, Wheeler lectured at a special meeting in Dallas, marking the discovery of quasars. *"Computer simulation shows that the collapse of a burnt-out star is remarkably similar to the highly idealized one computed by Oppenheimer and Snyder."*

As seen by an outside observer, the collapse slows down and becomes frozen at the critical radius. But as seen by an observer moving with the star's surface, the collapse is continuous right through the critical radius and on inward without hesitation.

Meanwhile in the hallway outside the lecture hall . . .

OPPIE, COME ON IN. WHEELER IS SHOUTING YOUR PRAISES ON COLLAPSING STARS. HE'S BEEN CON-VERTED!
TRAITOR
HIROSHIMA
COMMUNIST
NAGASAKI
PLEASE, DON'T BOTHER ME. CAN'T YOU SEE I'M MEDITATING ON THE BHAGAVAD GITA!

Wheeler was heartsick that Oppenheimer had lost interest in collapsing stars. But Oppie was worn out by years of political intrigue – directing the Manhattan Project, dealing with the tragedy of Hiroshima and Nagasaki, accusations of disloyalty to his country, and ignominiously losing his security clearance. Like a burnt-out star, the former **wunderkind** was himself collapsing into his own world, cut off from the rest of the Universe.

But for Wheeler, a new chapter in the history of physics had begun. *"Whatever the outcome of our studies, one feels that at last in stellar implosion we have a situation where general relativity dramatically comes into its own and where its fiery merge with quantum physics will be consummated."*

At that time, in 1962, Stephen William Hawking arrived at Cambridge University. Hawking was destined to take the first step in Wheeler's dream scenario of combining general relativity and quantum mechanics. But he was already feeling the symptoms of a disease that would, in ten years time, put him in a wheelchair and, in twenty years, destroy his ability to speak.

The Hawking Era

A visitor to the Cambridge University Department of Applied Mathematics and Theoretical Physics (DAMTP) would find a large photograph of the present Lucasian Professor of Mathematics, Stephen Hawking, permanently displayed in DAMTP's modest reception area beside the portraits of two giants of mathematical physics who had previously held this post, Sir Isaac Newton and Paul Dirac – world-renowned for his work on relativistic quantum mechanics.

An original copy of Hawking's thesis of 1965 is tucked away, with a few hundred others, in the first-floor library of DAMTP, with most of the equations written in his own hand. This manuscript represents the beginning of a new era in modern cosmology.

Hawking had come to Cambridge from Oxford to study with the world-famous cosmologist, Sir Fred Hoyle. But he was disappointed.

My application to do research at Cambridge was accepted, but, to my annoyance, my supervisor was not Hoyle but a man called Dennis Sciama, of whom I had not heard. Sciama, like Hoyle, believed in the steady state theory, according to which the universe had no beginning or end in time.
In the end, however, this turned out to be for the best. Hoyle was abroad a lot, and I probably wouldn't have seen much of him.
On the other hand, Sciama was there, and he was always stimulating, even though I often didn't agree with his ideas.

With his characteristic intellectual bravado, Hawking called his doctoral thesis, **Properties of Expanding Universes**. Already in the second line of the thesis abstract, there is a reminder of Hawking's early days at Cambridge. He summarized . . .

CHAPTER ONE SHOWS THAT THE EXPANSION OF THE UNIVERSE CAUSES GREAT DIFFICULTIES FOR THE HOYLE/ NARLIKAR THEORY OF GRAVITATION.

Fred Hoyle was the best-known of the three authors of the steady state theory of the Universe, along with Hermann Bondi and Thomas Gold, two refugees from Nazi Europe.

BBC
BBC
THIS MODEL PROPOSES THAT MATTER IS CONTINUOUSLY BEING CREATED AS THE UNIVERSE EXPANDS, IN UTTER DISAGREEMENT WITH THE BIG BANG NOTION OF AN INFINITELY DENSE INITIAL STATE.

In the early 60s, the steady state model was probably accepted by more astrophysicists and cosmologists than the big bang. Hoyle was particularly upset by aspects of the opposing model. On a BBC radio show in 1950, he had the ignominious distinction of being the first to call it the **Big Bang** – in derision, of course.

THIS INSTANTANEOUS CREATION OF THE UNIVERSE IS LIKE A PARTY GIRL JUMPING OUT OF A BIRTHDAY CAKE, IT'S RIDICULOUS, I CALL IT THE BIG BANG. NOW, ON THE OTHER HAND, MY OWN STEADY STATE THEORY...
BBC

Twelve years after this jibe, Hoyle was still developing aspects of gravitation theory at DAMTP, with a postgraduate student named Jayant Narlikar, to support the steady state model.

Hawking, who was floundering with his own research in his first months at Cambridge, became interested in Narlikar's calculations and began hanging around his office in the spirit of DAMTP's policy of free inquiry, open discussion and sharing of ideas. Hoyle knew nothing of this.

Hawking had become more and more involved in Narlikar's difficulties with the project Hoyle had assigned.

An experienced publicist, Hoyle would often present his ideas in advance of publication, before the work was refereed, in order to keep his name in the newspapers and the research grants coming in. He scheduled a talk at the prestigious Royal Society to discuss his latest ideas based on Narlikar's calculations.

Hawking stands up, clutching his stick, and the room falls silent.

Hoyle was furious, as an embarrassed laugh passed through the room. It was a dramatic confrontation between one of the world's best-known cosmologists and the student he had rejected. The session was quickly adjourned.

Hawking was indeed right about the divergence of Hoyle's equations, and the new approach was abandoned. Hoyle had essentially had his work "refereed" in an open forum by an unknown postgraduate student.

Later, Stephen wrote a paper summarizing the mathematical methods he had used, which established him as a promising young researcher.

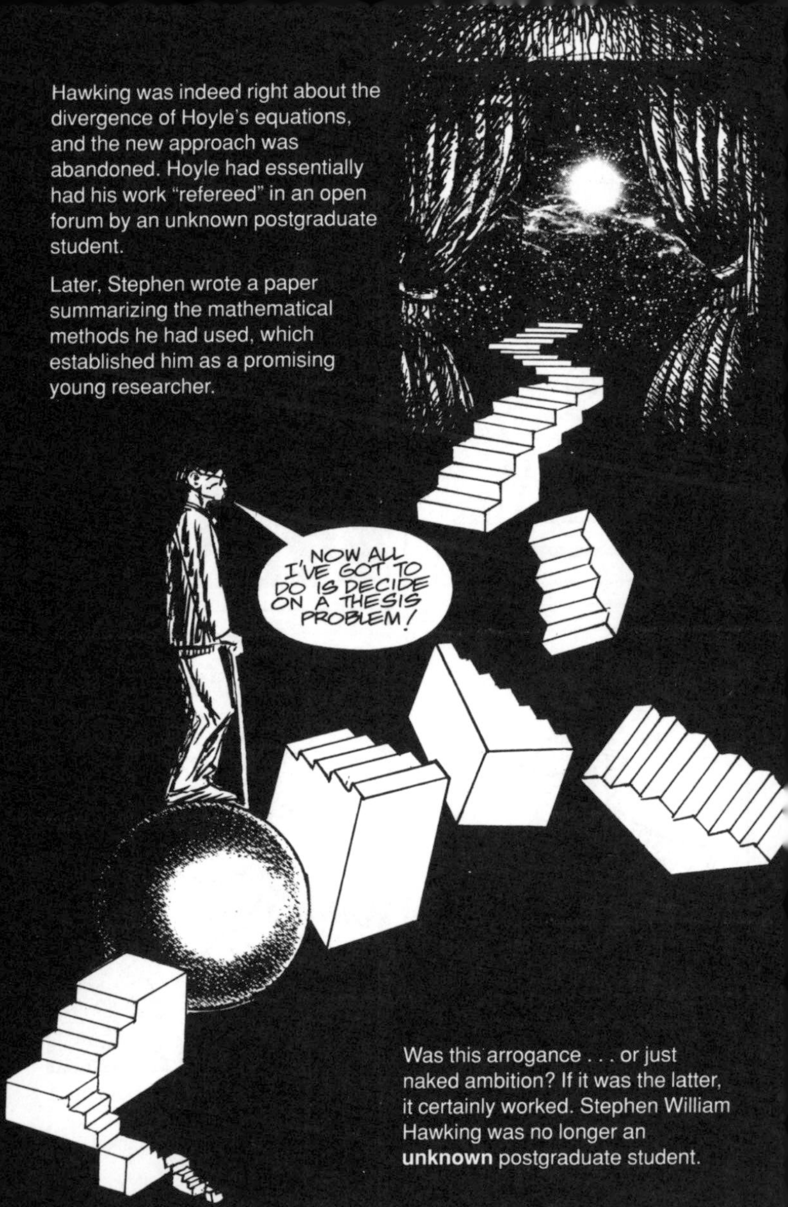

Was this arrogance . . . or just naked ambition? If it was the latter, it certainly worked. Stephen William Hawking was no longer an **unknown** postgraduate student.

The Unselfish Thesis Supervisor

Dennis Sciama turned out to be a committed thesis supervisor, in the true tradition of the unselfish tutor who stimulates his charges to look for ways to increase their experience.

He refused to speed up Hawking's doctoral programme, even when pressured by Stephen's persuasive father.

Sciama developed a unique style of supervising his postgraduate students. He would not share in their work, as many other professors did around the world (he has hardly ever written any joint papers). He does not even choose their topics.

If one wishes to study the big bang origin of the Universe with the cosmic radiation background, then cosmology is only understandable with general relativity. So, naturally when I set up a research school in Cambridge in the 1960s with students who seemed gifted enough to work in these difficult areas, I suggested general relativity.

Yet nearly all the students Sciama took on in those early days have had outstanding careers in cosmology:

- George Ellis is a Professor of Physics in South Africa. (Ellis wrote a book with Hawking entitled **Large Scale Structure of Space Time**, considered the Bible for research in relativistic cosmology. It is dedicated to D.W. Sciama.)
- Martin Rees, currently the Director of the Institute of Astronomy in Cambridge.
- Brandon Carter is currently Director of Research at the Observatory in Paris.

- And, of course, Stephen Hawking, Lucasian Professor at Cambridge.

One of Sciama's important activities was to arrange for his students to attend important seminars. He always seemed to know what was going on. In the mid 60s, the Cambridge group became interested in the work of a young applied mathematician, Roger Penrose, then based at Birkbeck College in London.

After graduating from Cambridge and research in the US, Penrose had begun to develop ideas about **singularity theory**, which matched well with the ideas of the Cambridge research group.

ALTHOUGH PENROSE WASN'T ONE OF MY FORMER STUDENTS, I DID SEDUCE THE PROLIFIC MATHEMATICIAN INTO WORK IN COSMOLOGY DURING THE LATE 1950S...
COFFEE HOUSE
100 YDS.

It was only a few years after John Wheeler had accepted Oppenheimer's solution and the existence of black holes, that Sciama started sharing his enthusiasm with some of his colleagues and students. Penrose, already one of the world's top mathematicians, got a flash of inspiration about these exotic objects from Sciama in a Cambridge coffee shop.

DENNIS, I'M SURE I CAN APPLY MY NEW MATHEMATICAL METHODS IN TOPOLOGY TO THE PROBLEMS OF COLLAPSING STARS...

Penrose was soon able to show that if a star collapses beyond a certain point, it could not re-expand. Within the framework of general relativity, the star could not avoid becoming infinitely dense, i.e. it would form a **singularity** at its centre.

It was **not** true – as many insisted – that the matter of that star would "fly past itself" and expand again. Instead, a singularity of space–time would occur, a point at which time came to an end and the laws of physics broke down. It was the first **singularity theorem**.

MATTER FLYING PAST ITSELF INSIDE A COLLAPSING STAR AND RE-EXPANDING – NOT TRUE, SAID PENROSE.

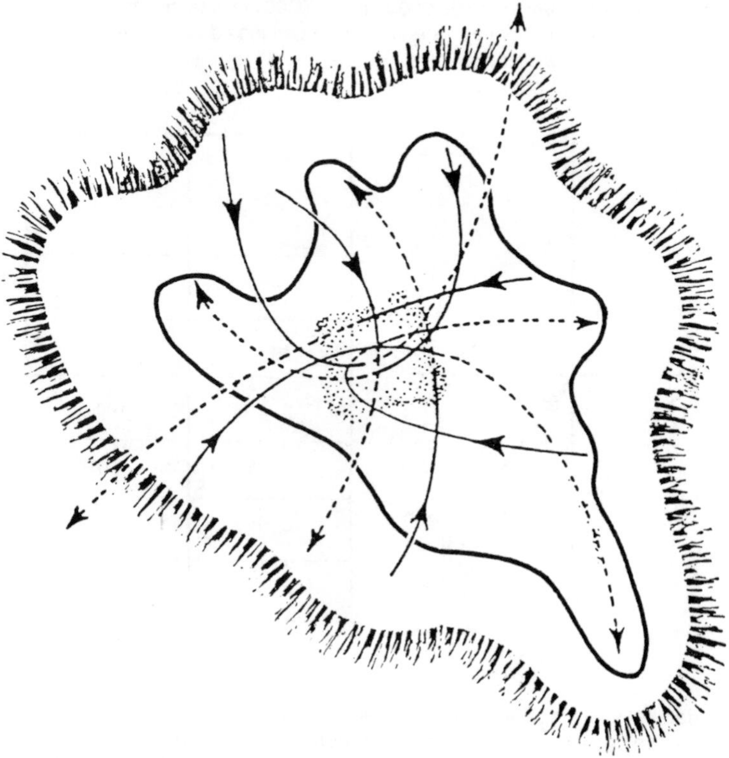

Something You Need to Know: What is a Singularity?

Generally speaking, a singularity is a point at which a mathematical function cannot be defined. The function is seen to diverge to infinitely large values.

For example, the simple algebraic equation Y = 1/X has a singularity for value X = 0. If we make positive values of X arbitrarily small, then Y is arbitrarily large in the vertical (or positive) direction.

If we then plug in arbitrarily small negative values of X, we find Y has an arbitrarily large negative value. Thus, for the smallest change imaginable in the variable X, say from +0.000001 to -0.000001, Y changes from +1 million to -1 million. Clearly at X equals 0, something has gone wrong. This is a mathematical singularity.

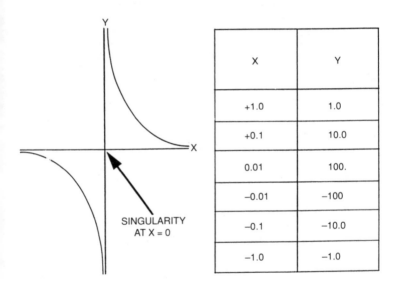

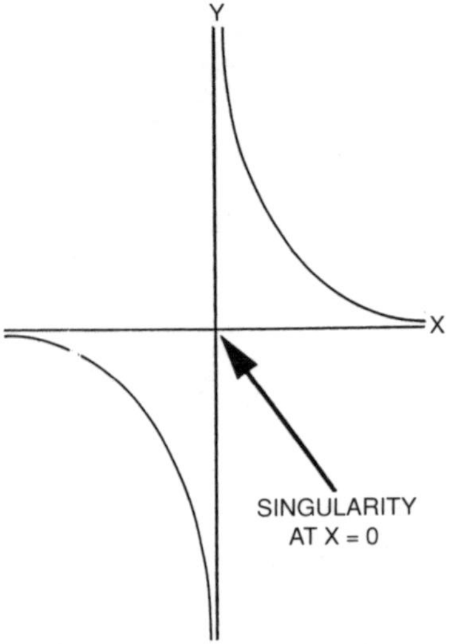

X	Y
+1.0	1.0
+0.1	10.0
0.01	100.
–0.01	–100
–0.1	–10.0
–1.0	–1.0

In general relativity, a singularity is a region of space–time in which the curvature becomes so strong that the general relativistic laws break down, and presumably the laws of quantum gravity take over.

If an attempt is made to describe a singularity using general relativity alone, an incorrect result is obtained: mainly that the curvature and the tidal gravity is infinite at that point. Quantum gravity probably replaces these infinities with "quantum foam" – and merges with the laws of general relativity.

But this does not mean that singularity points cannot be studied and the physics near these points understood. There are certain singularity theorems that yield important qualitative information under certain conditions. For example, if the mathematics are handled carefully, the proof of the existence of a true singularity can be a result with physical meaning. Thus, the singularity theorems of Penrose and, later, Hawking.

In the Schwarzschild solution of Einstein's field equations, the critical radius is not a real singularity (in spite of its early description as the Schwarzschild singularity). The physical processes are continuous across the boundary, and a simple change in the mathematical coordinates removes the divergence.

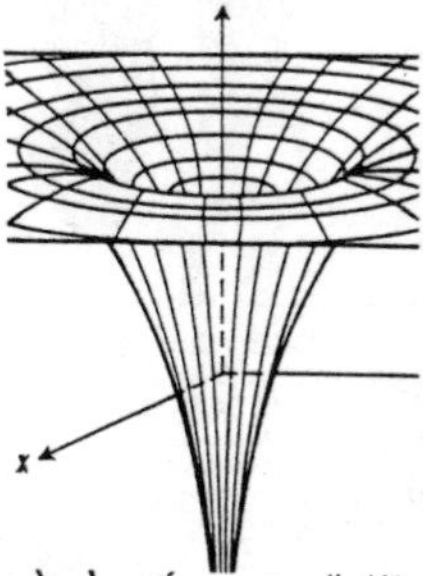

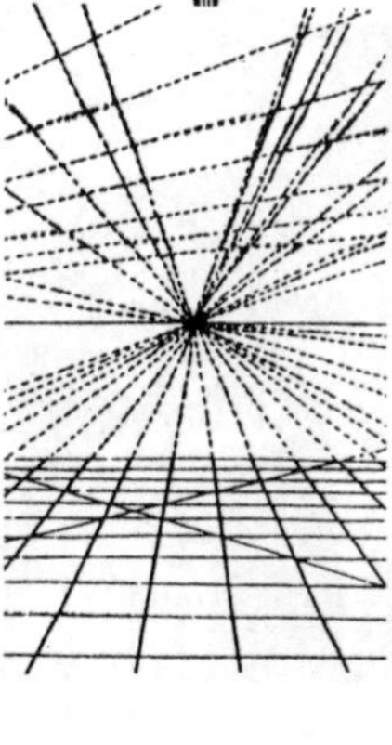

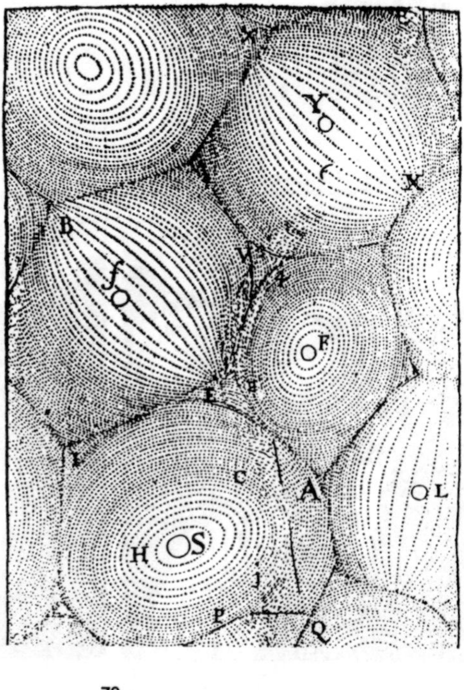

SEMINAR TODAY
ROGER PENROSE:
Topological Methods in Singularity Theorems

A group of Sciama's students were at Penrose's London seminar when he announced that he had proved that **a singularity definitely exists when a star collapses to form a black hole**.

Stephen Hawking was not at Penrose's seminar that day. But the news reached him immediately and made a deep impression.

SEMINAR TODAY
ROGER PENROSE:
Proof of Black Hole Singularities

PENROSE'S RESULTS ARE VERY INTERESTING. I WONDER IF THEY COULD BE ADAPTED TO UNDERSTANDING THE ORIGIN OF THE UNIVERSE; THE EXPANDING UNIVERSE AS A GIANT COLLAPSING STAR IN REVERSE.
YOU MEAN BY REVERSING THE SENSE OF TIME...
YES. MAYBE THE SAME CONSIDERATION APPLIES AS IN HIS THEOREM FOR STARS. I'M GOING TO TRY TO ADAPT HIS RESULT TO THE WHOLE UNIVERSE AND SEE WHAT HAPPENS.
OKAY. IT SHOULD BE VERY INTERESTING.

Hawking had just one year left as a research student, and only now did he have a challenging problem. To adapt Penrose's method, he had to work hard, learn the mathematics involved and write it up as the last chapter in his thesis – his first singularity theorem for the **beginning of the Universe**.

Hawking had shown that if general relativity is correct, there must have been a singularity in the past which was a beginning of time.

Hawking passed and received his PhD in 1965. There were some complications – like infinite and non-infinite universes – but over the next few years he developed new techniques to remove these problems.

This has become generally accepted and today everyone assumes that the Universe started with a big bang – a highly dense and hot initial state. This is Hawking's major contribution to big bang cosmology, a major result for which he was to become known worldwide. Thus by 1970, five years after receiving his PhD, Stephen Hawking was an internationally known cosmologist.

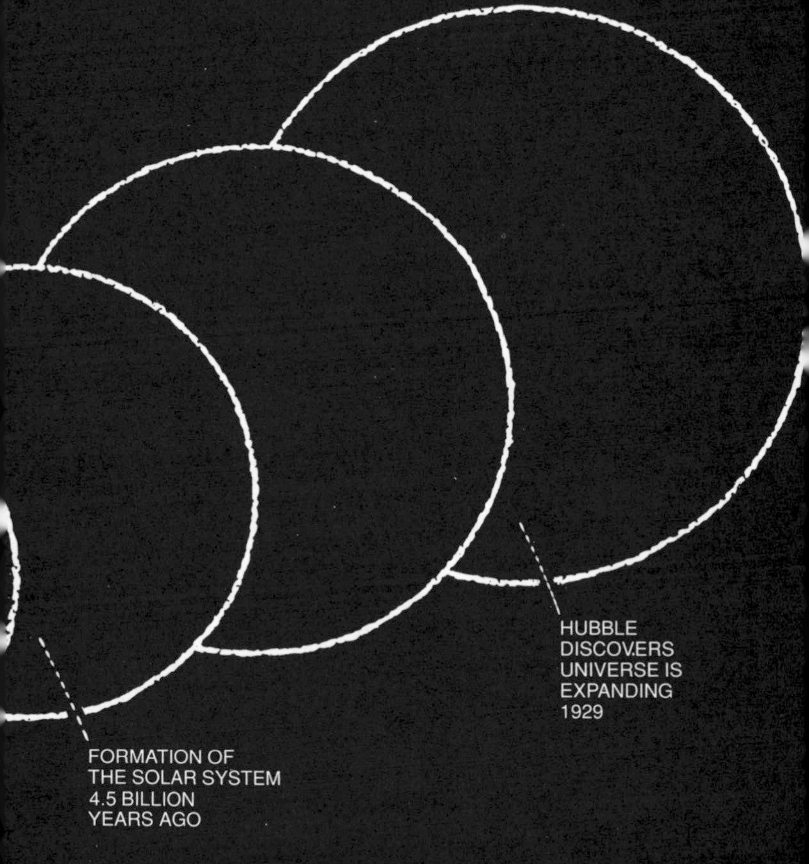

Stephen Hawking has been a proponent of the big bang model since his early days as a postgraduate student. His PhD thesis which criticized Hoyle's steady state model and his proof of a big bang singularity link his name with the success of the latter for all time.

It is interesting to imagine the recent history of cosmology (or at least the recent history of Stephen Hawking), if his application to study with Hoyle at Cambridge had been approved.

Today, Hoyle and his former student of 30 years ago, Jay Narlikar, are still patching up the steady state model. But it is a dead duck. The world of cosmology has moved on. Perhaps this is best shown by the **Scientific American** article in the October 1994 special issue on the Universe, which promises to become the accepted description of our understanding of the Universe into the next millennium.

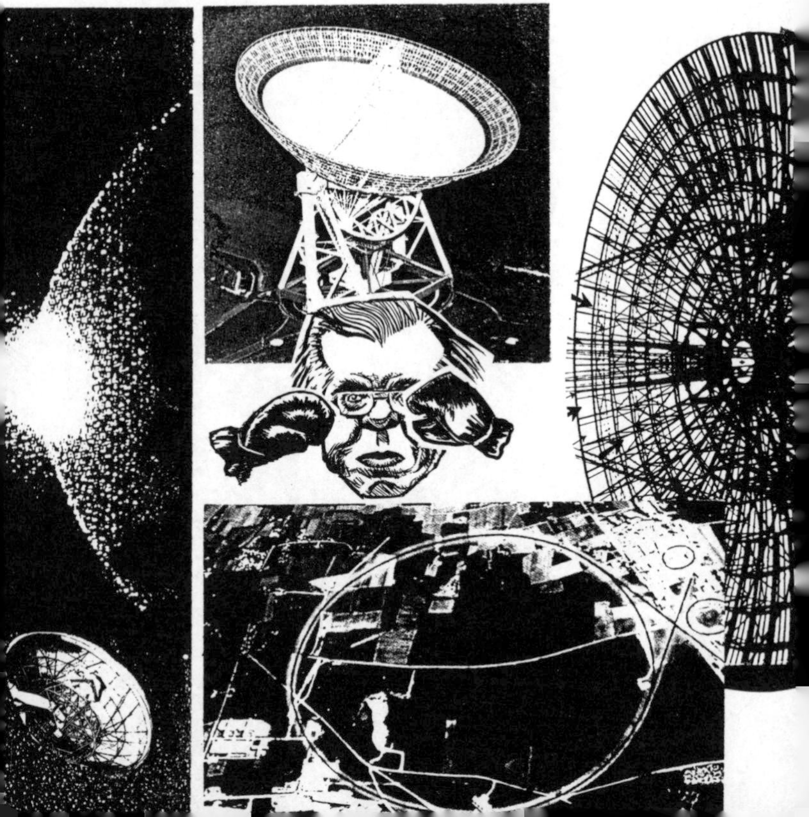

The Evolution of the Universe

Understanding of the evolution of the Universe is one of the great achievements of 20th-century science. This knowledge comes from decades of innovative experiments and theories. Modern telescopes on the ground and in space detect the light from galaxies billions of light-years away, showing us what the Universe looked like when it was young. Particle accelerators probe the basic physics of the high energy environment of the early Universe. Satellites detect the cosmic background radiation left over from the early stages of expansion, providing an image of the Universe on the largest scales we can observe.

Our best efforts to explain this wealth of data are embodied in a theory known as the standard cosmological model or the big bang cosmology. The major claim of the theory is that in the large-scale average the Universe is expanding in a nearly homogeneous way from a dense early state.

*At present, there are no fundamental challenges to the big bang theory, although there are certainly unresolved issues within the theory itself. Astronomers are not sure, for example, how the galaxies were formed, but there is no reason to think the process did not occur within the framework of the big bang. Indeed, the predictions of the theory have survived all tests to date. (**Scientific American**, October 1994).*

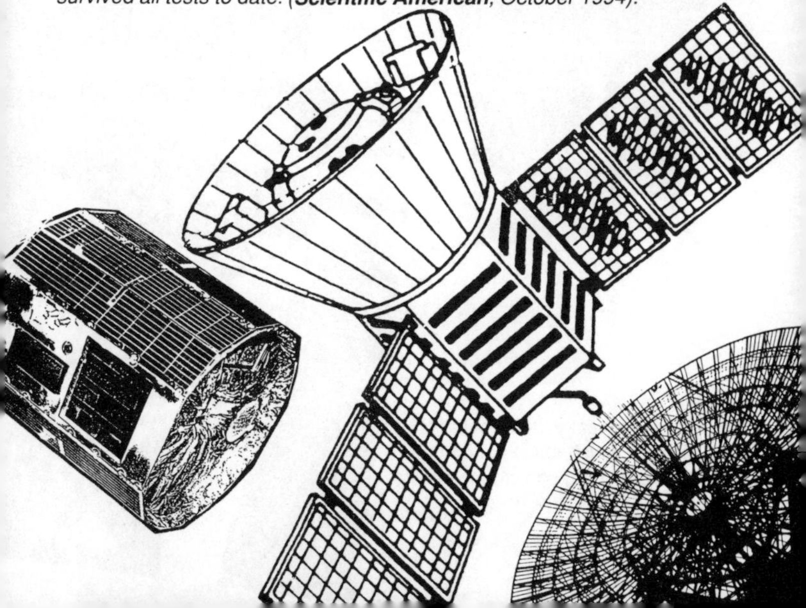

1965: a Big Year for Hawking

Hawking married his sweetheart Jane Wilde in Trinity Chapel at Cambridge in July 1965. Though he was now hobbling more and more on his cane, he had his PhD, a devoted and intelligent wife and new mathematical skills to use in cosmology. He also received a fellowship at Caius College to continue work at DAMTP. He was no longer depressed.

And still that cocky, determined look on his face which said . . . I can do anything. Nothing can stop me, not even ALS.

An Unstoppable Mind

Stories abound of Hawking's prodigious mental abilities, already apparent in his Oxford undergraduate years.

Several fellow students had spent weeks on a major assignment, some thirteen problems from a difficult text, *Electricity and Magnetism* by Bleaney & Bleaney. They were told to do as many as possible. Most managed to complete only one or two in the time allotted. Characteristically, Hawking left it to the last day. After spending the morning in his room, he emerged to say he was only able to complete the first **ten** of the problems!

One of the Oxford tutors supervising Hawking's work in statistical physics had assigned several problems from a textbook which Stephen disliked. At the next tutorial he returned, not with the work completed, but with all the mistakes in the textbook marked out. The tutor quickly realized that Hawking knew more about the subject than he did.

Near the end of his term at Oxford and no doubt beginning to feel the effects of ALS, Hawking took a terrible fall down a staircase in the university hall. As a result, he temporarily lost his memory. He could not even remember his name.

After several hours of interrogation by his friends, he finally returned to normal but was worried about possible permanent brain damage. To be sure, he decided to take the Mensa test for individuals with superior intelligence. He was delighted to find that he had passed with flying colours, scoring between 200 and 250!

Nothing, not even the dreadful illness of ALS, could stop that mind.

The Sixties Revolution

It is debatable whether or not social historians of the 21st century will continue to analyse and report the decade of the 1960s as a period of great social upheaval and radical change on the Earth.

But it is certain that historians of science will view the same period as a time of radical change in our understanding of the cosmos. It is already being referred to as the golden age of relativistic cosmology – and **relativistic cosmology** is where it's at!

Heroes of the 60s – from the moppet-headed Beatles to the crowd at Woodstock – have become familiar icons. Similarly, the revolution in cosmology also has its heroes, but they are mostly unknown to the general public.

The 60s was a time of remarkable progress in observational astronomy, due mainly to major advances in technology and instrumentation. All sorts of observed unusual phenomena led to new models of celestial objects, which can only be described as a revolution in cosmology. The beginning of this revolution can be traced to a crucial meeting at a time and place indelibly marked in the history of the 20th century – but for a different reason.

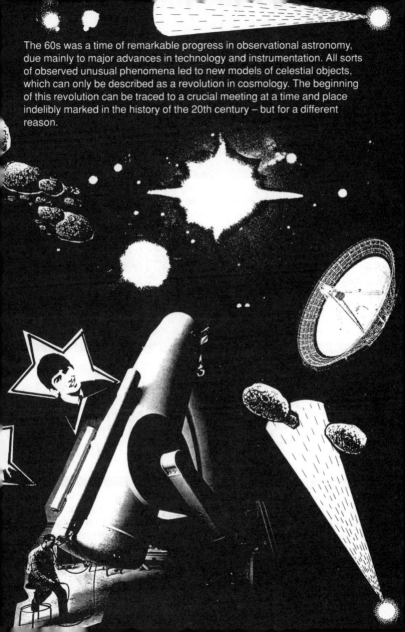

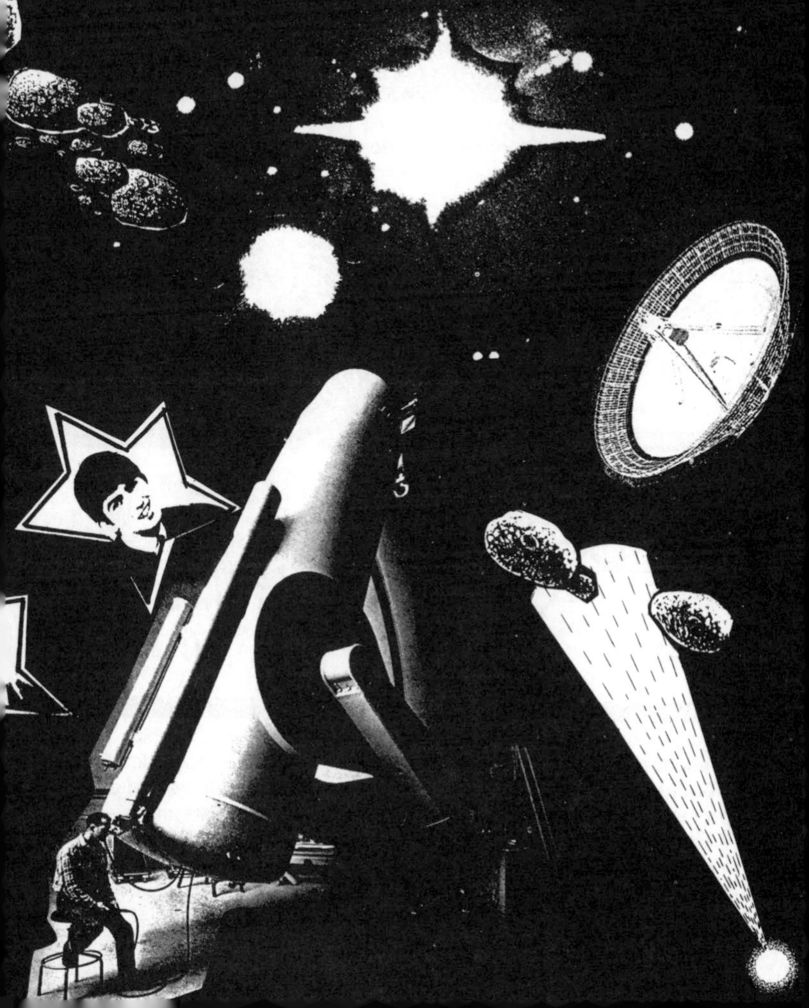

Dallas 1963

If you ask a sample of over people over fifty years old if they remember **Dallas 1963**, most will immediately describe exactly what they were doing when John F. Kennedy was gunned down in Dallas on 22 November.

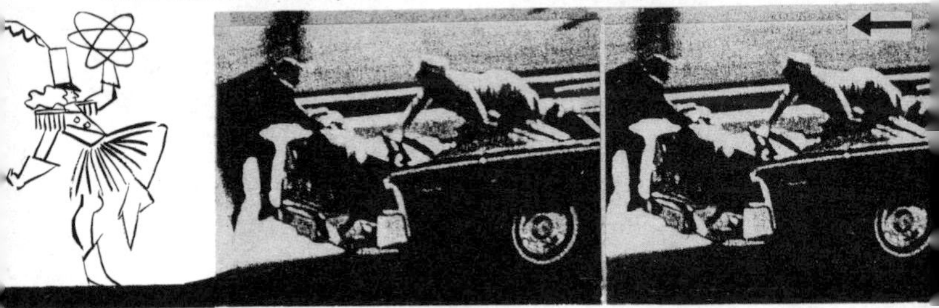

But perhaps one small subset of that group might give an ambiguous response. Of course, they would remember Kennedy's tragic death. But **Dallas 1963** would have another *special* connotation for the group of three hundred astronomers, astrophysicists, cosmologists and relativists who attended the **First Texas Symposium on Relativistic Astrophysics** to mark the discovery of quasars. The symposium was held in Dallas, 16–18 December 1963, only three weeks after JFK's assassination.

The **relativists**, odd-ball specialists who spent their working lives playing around with Einstein's equations, had been invited to join real astronomers and astrophysicists in a dialogue. At last, 25 years after the famous Oppenheimer and Snyder paper on collapsing stars, general relativity was being suggested as a possible explanation for a new physical phenomenon that **had actually been observed by practical working astronomers**.

It was thought that gravitationally collapsed stars (soon to be called black holes) might be producing the massive energy necessary to explain observations on the new and exciting objects called quasars.

Thomas Gold, one of the developers of the Steady State Universe, gave the after-dinner speech at the Dallas Symposium.

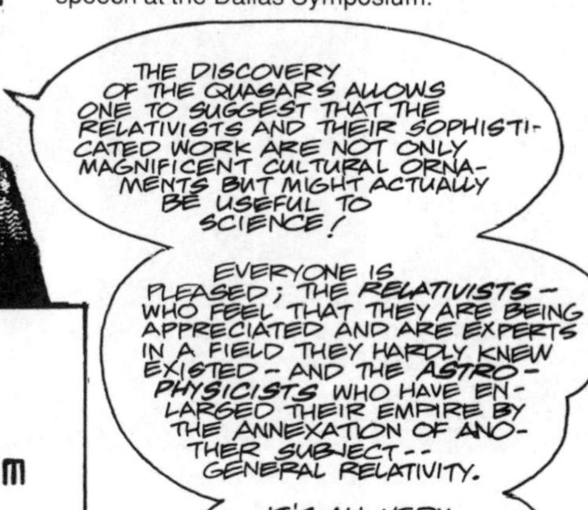

THE DISCOVERY OF THE QUASARS ALLOWS ONE TO SUGGEST THAT THE RELATIVISTS AND THEIR SOPHISTICATED WORK ARE NOT ONLY MAGNIFICENT CULTURAL ORNAMENTS BUT MIGHT ACTUALLY BE USEFUL TO SCIENCE!
EVERYONE IS PLEASED; THE RELATIVISTS – WHO FEEL THAT THEY ARE BEING APPRECIATED AND ARE EXPERTS IN A FIELD THEY HARDLY KNEW EXISTED – AND THE ASTROPHYSICISTS WHO HAVE ENLARGED THEIR EMPIRE BY THE ANNEXATION OF ANOTHER SUBJECT -- GENERAL RELATIVITY.
IT'S ALL VERY PLEASING, SO LET US ALL HOPE THAT IT IS RIGHT.

It did turn out to be right, as Hawking himself modestly admits 30 years later.

There has been a great change in the status of general relativity and cosmology in the last thirty years. When I began research in the Department of Applied Mathematics and Theoretical Physics (DAMTP) at Cambridge in 1962, general relativity was regarded as a beautiful but impossibly complicated theory that had practically no contact with the real world. Cosmology was thought of as a pseudo-science where wild speculation was unconstrained by any possible observations.

The position today is very different, partly due to the great expansion in the range of observations made possible by modern technology, but also because we have made tremendous progress on the theoretical side.

This is where I can claim to have made a modest contribution.

But observations on quasars required completely new observational techniques. So before describing the excitement about quasars, it might be a good idea to review ***something you need to know.***

Something You Need to Know: the Electro Magnetic Spectrum

The **electromagnetic spectrum** sounds very technical because the two words are seldom used outside physical science. The first term, *electromagnetic,* just means that the waves we will speak of (light, radio, infrared) are made up of vibrating electric and magnetic fields. The second term, *spectrum,* refers to the range of sizes of the waves, i.e. their wavelengths.

The EM spectrum refers to all the possible wavelengths of radiation existing in nature. Different-sized waves have different properties and are generated by different physical processes. Furthermore, they must be detected by completely different equipment. The invisible radiation coming from stars and galaxies (in addition to the visible or optical band) gives useful information, though it can't be seen with the unaided eye.

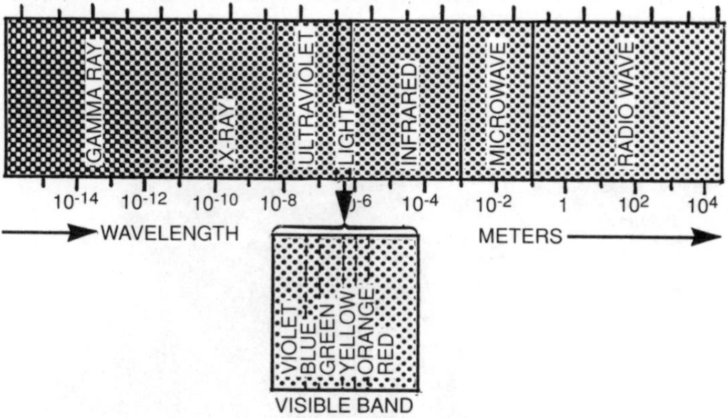

The wavelengths cover a wide range of values from X-rays (smaller than the distance between atoms) to radio waves (several kilometres in length). The waves all travel at the same speed as light and there is a remarkably simple relationship between the wavelength, the frequency of the source emitting the waves and the speed of transmission:

(**wavelength) x (frequency) = (speed of light)**

Before the 1960s, observational astronomy meant optical (or visible) astronomy only – looking through telescopes composed of glass lenses or reflecting mirrors and observing with the eye or with very sensitive cameras. Special films did extend measurements into the invisible infrared band with longer wavelengths than visible light.

But during the late 1950s and 1960s, nearly the **whole** electromagnetic band became detectable to observational astronomers, such that now we have *radio* astronomy, *microwave* astronomy, *infrared* astronomy, *optical* astronomy, *ultraviolet* astronomy, *X-ray* astronomy and even *gamma-ray* astronomy.

The great discoveries of the 1960s came from these extensions of observations outside the visible range, particularly to the longer wavelength microwave and radio bands. **Quasars** and **pulsars** were discovered in the **radio frequency band** and the **cosmic background radiation** was detected in the **microwave band**. And in the 1970s, **X-ray astronomy**, at the other end of the spectrum, produced the first evidence for the existence of black holes from observations of the constellation Cygnus X-1.

MY ELECTRO-MAGNETIC THEORY, IN 1867, PREDICTED THE EXISTENCE OF ALL THESE WAVES.
JAMES CLERK MAXWELL... ANOTHER CAMBRIDGE MAN.

1963: Quasars

Careful observations by radio and optical astronomers in the years 1960 to 1962 showed that there were over a half-dozen bright objects in the sky which were small enough to be stars but had a weird light spectrum – not like any star seen before.

Everyone was puzzled until 5 February 1963 when astronomers Maartin Schmidt and Jesse Greenstein at Caltech made a discovery.

Measurements indicated that these *quasi-stellar objects* (later to be named *quasars*) were moving away from the Earth at enormous speeds and therefore must be very, very far away.

They were first thought to be stars in the Milky Way galaxy, but their discoverers soon argued that these objects were moving away from the Earth as a result of the Universe's expansion. At the enormous distances calculated, their brightness implied they were radiating 100 times more energy than the **most luminous galaxy ever seen**.

QUASARS. LIGHT LEAVES THE QUASAR AT POINT **A**. BILLIONS OF YEARS LATER, AT POINT **B**, THE LIGHT HAS STILL NOT REACHED THE MILKY WAY. WHEN THE LIGHT FINALLY REACHES US AT POINT **C**, WE DETECT IT AS IT WAS ALL THE WAY BACK AT POINT **A**.

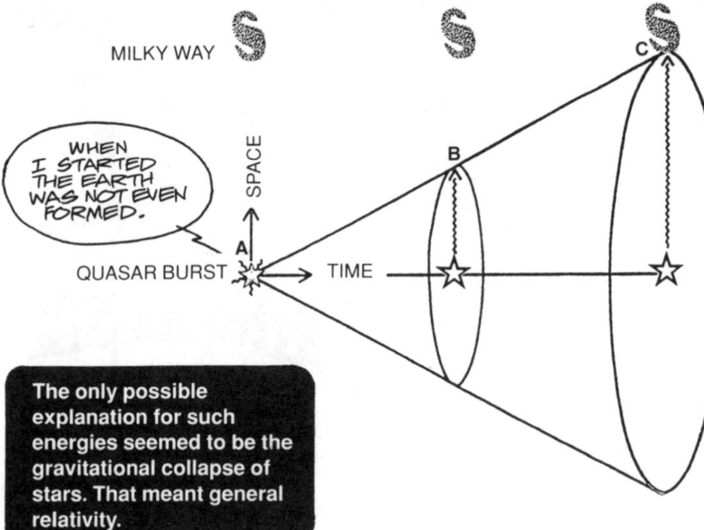

The only possible explanation for such energies seemed to be the gravitational collapse of stars. That meant general relativity.

1965: the Cosmic Background Radiation

In 1965, an accidental discovery of mysterious microwaves from outer space turned out to be the first experimental indication that the Big Bang model might be correct. Until that time the model was thought to be something of a joke. Here's how it happened . . .

The picture of the Universe as a primordial atom ("cosmic egg") by Abbé Georges Lemaître in 1927 led some cosmologists to picture the early universe as a hot, dense, rapidly evolving plasma. One of the more imaginative of these theorists, a free-thinking Russian émigré to the USA named George Gamow, considered the effect of the cooling of this plasma as the Universe expanded.

He then made one of the most important predictions in the history of science.

THE UNIVERSE MAY TODAY BE FILLED WITH A COSMIC BACKGROUND RADIATION COMPOSED OF ANCIENT PHOTONS RELEASED BY THE BIG BANG.
AFTER CORRECTIONS OF SOME ERRORS IN MY ORIGINAL CALCULATION, IT WAS SHOWN THAT THE TEMPERATURE OF THIS RADIATION SHOULD TODAY BE ABOUT FIVE DEGREES ABOVE ABSOLUTE ZERO.

Two of Gamow's colleagues, Ralph Alpher and Robert Herman, actually predicted that this radiation could still be detected.

All hot bodies (i.e. any object which has a temperature) give off continuous electromagnetic waves called thermal radiation, **even** if the temperature is only 5 degrees above absolute zero. The question was how to measure the radiation – which wavelength band to search.

To follow this part of the story, there is ***Something Else You Really Need to Know!***

Something You Need to Know — Thermal Radiation

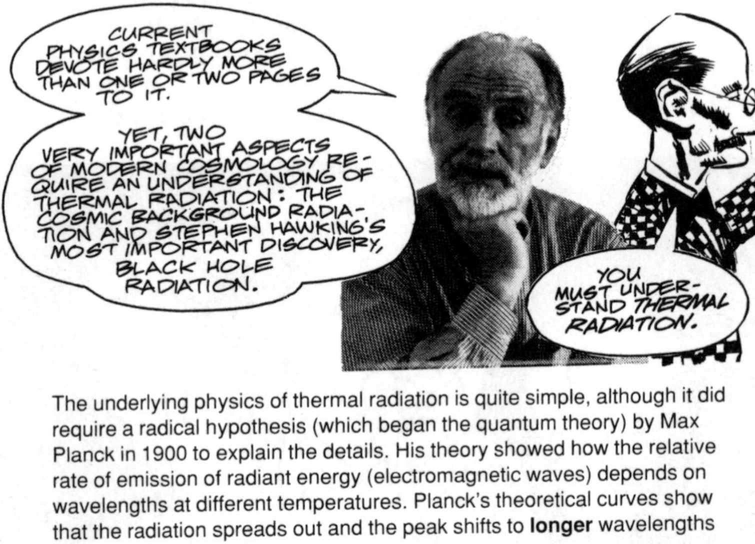

CURRENT PHYSICS TEXTBOOKS DEVOTE HARDLY MORE THAN ONE OR TWO PAGES TO IT.
YET, TWO VERY IMPORTANT ASPECTS OF MODERN COSMOLOGY RE-QUIRE AN UNDERSTANDING OF THERMAL RADIATION: THE COSMIC BACKGROUND RADIA-TION AND STEPHEN HAWKING'S MOST IMPORTANT DISCOVERY, BLACK HOLE RADIATION.
YOU MUST UNDER-STAND THERMAL RADIATION.

The underlying physics of thermal radiation is quite simple, although it did require a radical hypothesis (which began the quantum theory) by Max Planck in 1900 to explain the details. His theory showed how the relative rate of emission of radiant energy (electromagnetic waves) depends on wavelengths at different temperatures. Planck's theoretical curves show that the radiation spreads out and the peak shifts to **longer** wavelengths as the temperature **drops**.

- At 800 degrees centigrade, enough visible radiation is emitted to appear red hot, though most of the energy emitted is in the infrared band.
- At 300 degrees centigrade practically all of the energy emitted is carried by waves longer than red light and are called infrared, meaning *beyond the red*. No radiation is emitted in the visible band.
- At 5 degrees above absolute zero (or minus 268 degrees centigrade) the radiation is completely beyond the infrared in the microwave band and special microwave receivers are required to make the measurements.

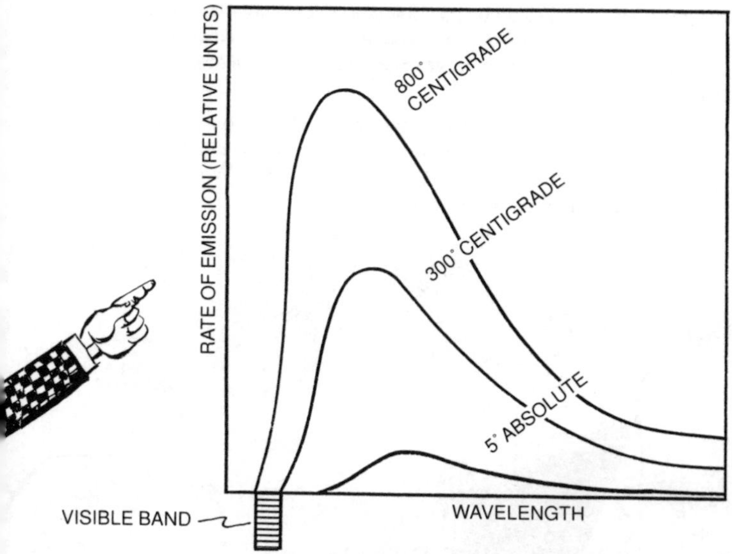

Since the shape of the curve is uniquely determined by the temperature of the emitting bodies, measurements at different wavelengths can infer the temperature of the body emitting the radiation. Conversely, if the temperature of the emitting body is known, then the shape and the distribution of the radiation can be predicted from theoretical formulae.

KEEP THIS INFORMATION IN MIND.
THIS SIMPLE PHYSICS IS CRUCIAL TO UNDERSTANDING THE RADIATION FROM THE COSMIC BACK-GROUND AND FROM BLACK HOLES.

Returning to Gamow's prediction, the theoretical curve for the *thermal radiation distribution* at 5 degrees above absolute zero indicated that the peak radiation should be in the **microwave** region of the electromagnetic spectrum.

While other groups were in the process of planning experiments to look for Gamow's microwaves, they were discovered accidentally by two researchers, Arno Penzias and Robert Wilson, at Bell Telephone Laboratories in northern New Jersey, USA.

DON'T BLAME US!
ROBERT, WHAT'S THAT PERSISTENT HISS WE'RE PICKING UP ON OUR MICROWAVE HORN ANTENNA?
YEAH, ARNO, IT'S BUILT FOR SATELLITE COMMUNICATION. MUST BE THOSE DAMN PIGEONS!

HISTORY OF THE UNIVERSE

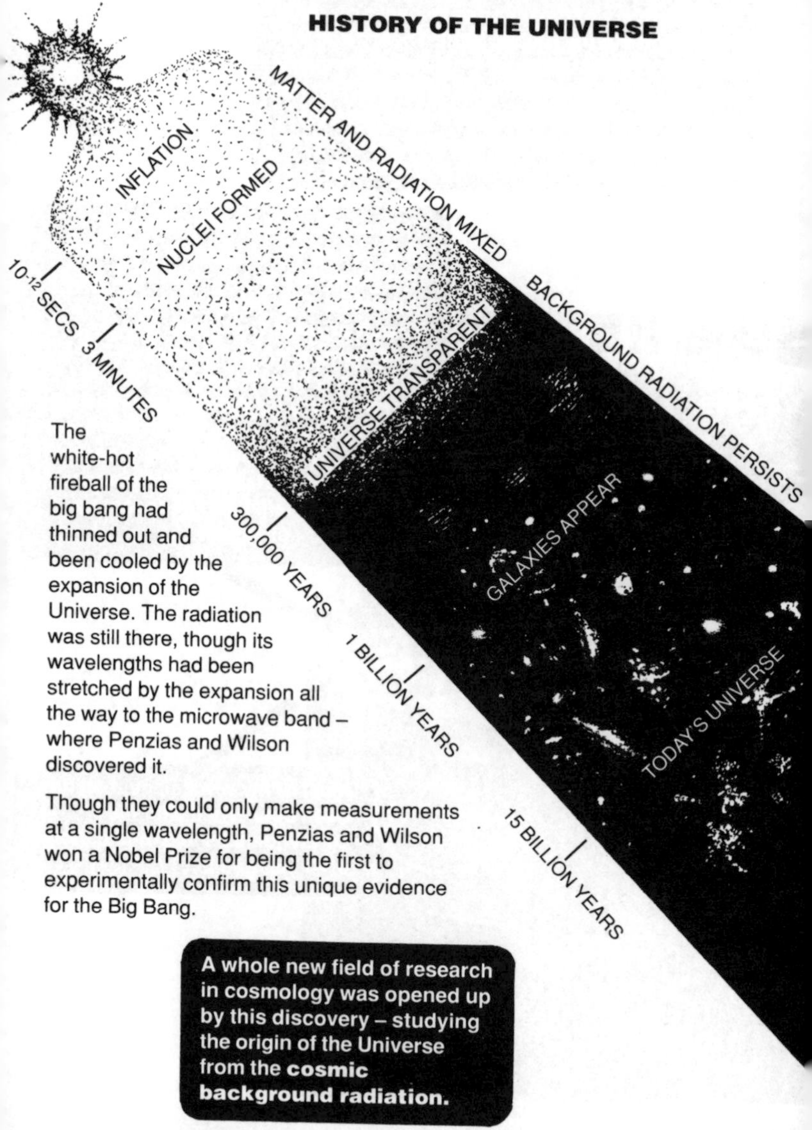

The white-hot fireball of the big bang had thinned out and been cooled by the expansion of the Universe. The radiation was still there, though its wavelengths had been stretched by the expansion all the way to the microwave band – where Penzias and Wilson discovered it.

Though they could only make measurements at a single wavelength, Penzias and Wilson won a Nobel Prize for being the first to experimentally confirm this unique evidence for the Big Bang.

A whole new field of research in cosmology was opened up by this discovery – studying the origin of the Universe from the cosmic background radiation.

The discovery of the microwave background in 1965 ruled out the Steady State Theory and showed that the Universe must have been very hot and dense at some time in the past. But the observations themselves did not exclude the possibility that the Universe bounced at some fairly large but not extremely high density.

This was ruled out on theoretical grounds by the singularity theorems that Penrose and I proved. We published **The Singularities of Gravitational Collapse and Cosmology**, an all-purpose singularity theorem which showed that the classical concept of time must have a beginning at a singularity in the past (i.e. the Big Bang). This theorem also implied that time would come to an end for at least part of spacetime when a star collapsed.

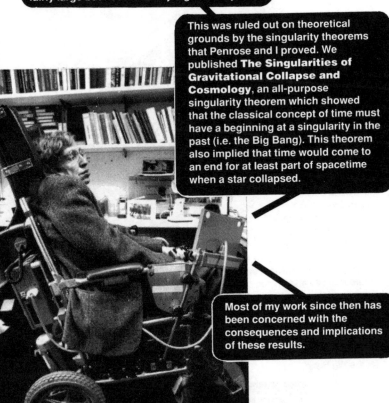

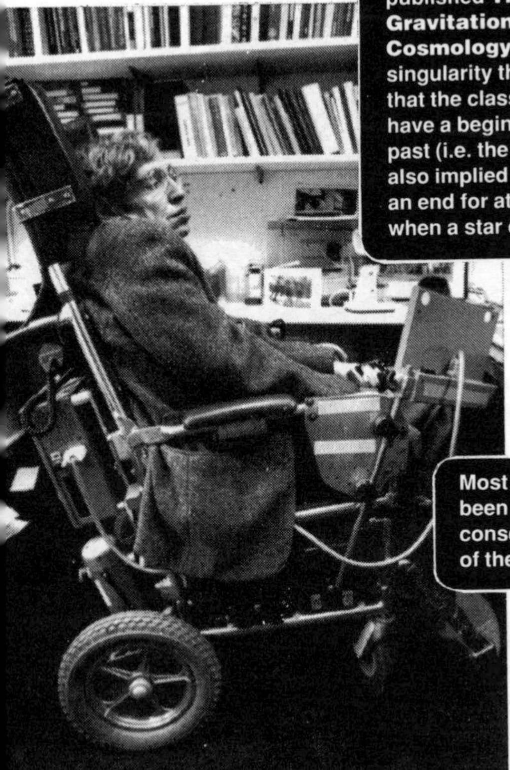

Most of my work since then has been concerned with the consequences and implications of these results.

Radio astronomers continued finding many more radio galaxies (i.e. galaxies emitting electromagnetic waves primarily in the radio waveband).

Then in 1967, a Cambridge postgraduate student named Jocelyn Bell detected highly regular sharp pulses at 3.7 metres wavelength from one of these galaxies. The Cambridge radio astronomers thought they had contacted an extra-terrestrial civilization!

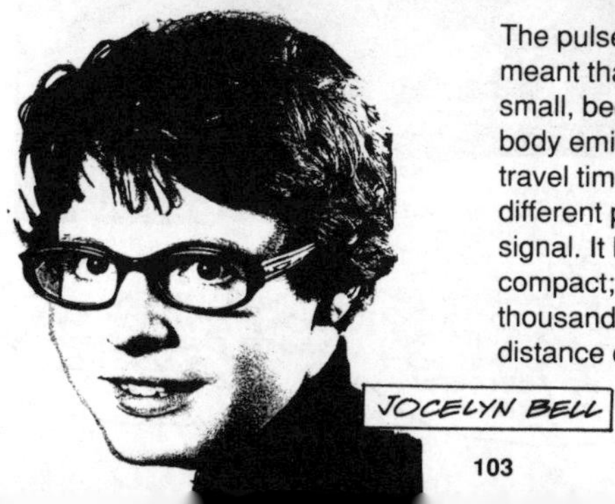

The pulses were very narrow. This meant that the emitter had to be very small, because you can't have a large body emitting short, sharp pulses. The travel time of the radiation from its different parts would smear out the signal. It had to be something highly compact; an object smaller than a few thousand kilometres in size, yet at the distance of a star.

As the Cambridge radio astronomers announced their results, the DAMTP theorists, Sciama, Hawking, Rees sat smugly at the seminar.

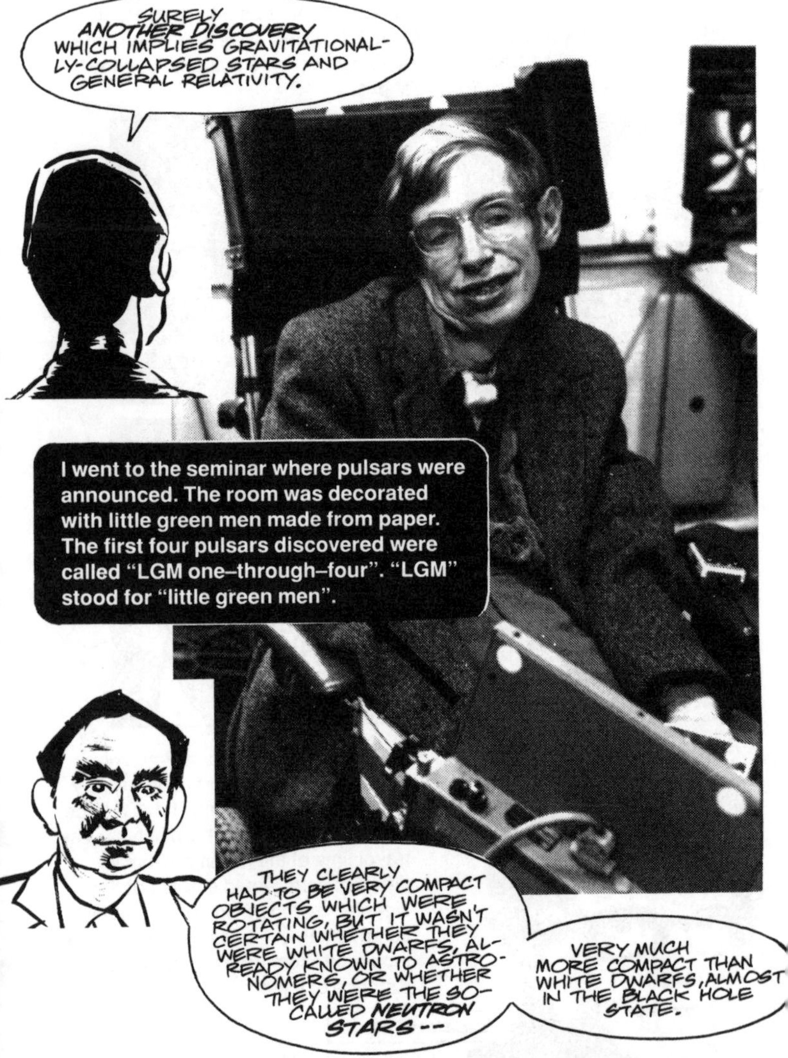

SURELY ANOTHER DISCOVERY WHICH IMPLIES GRAVITATIONALLY-COLLAPSED STARS AND GENERAL RELATIVITY.
I went to the seminar where pulsars were announced. The room was decorated with little green men made from paper. The first four pulsars discovered were called "LGM one–through–four". "LGM" stood for "little green men".
THEY CLEARLY HAD TO BE VERY COMPACT OBJECTS WHICH WERE ROTATING, BUT IT WASN'T CERTAIN WHETHER THEY WERE WHITE DWARFS, ALREADY KNOWN TO ASTRONOMERS, OR WHETHER THEY WERE THE SO-CALLED NEUTRON STARS --
VERY MUCH MORE COMPACT THAN WHITE DWARFS, ALMOST IN THE BLACK HOLE STATE.

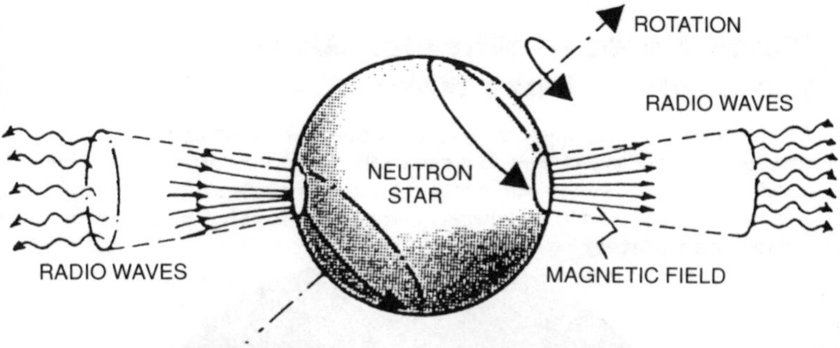

It took a few months of discussion before it became clear. Tommy Gold, who had worked earlier on the steady state theory, was the first to make the argument clear.

PULSARS ARE ROTATING NEUTRON STARS AND CAN'T BE ANYTHING ELSE. RADIO WAVES BEAMING OUT OF THE NEUTRON STAR REACH THE EARTH INTERMITTENTLY AS THE STAR ROTATES, SIMILAR TO A LIGHTHOUSE.

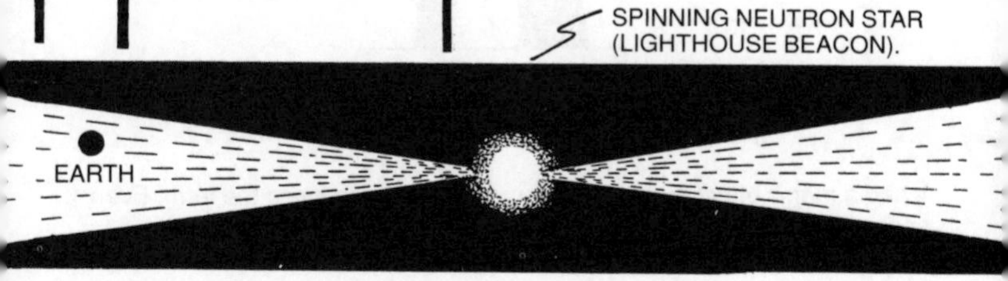

SPINNING NEUTRON STAR (LIGHTHOUSE BEACON).
EARTH

Black Holes — Wheeler Gives the Media a Buzz Word

As the 1960s were coming to a close, everyone was talking about *gravitationally collapsed stars*. The **partially** collapsed stars – white dwarfs and neutron stars – had become everyday objects to astronomers. But John Wheeler was interested in **massive stars** which collapse **completely**.

It had a magic effect. Everybody immediately began using the term. Even specialists could now know they were speaking about the same thing. In Moscow, Pasadena, Princeton and Cambridge, black holes replaced "gravitationally completely collapsed stars".

The Age of Black Holes

The media went nuts. At least they could encapsulate all this new complicated physics and astronomy in two simple words which fell easily into newspaper columns. Writers picked up on the new buzz word and books appeared on the popular science and sci-fi shelves. On TV, *Star Trek* had exotic new destinations for its space ships. At dinner parties, scientists were put on the spot to explain black holes to friends. Black holes had become household words . . . but did anyone really know what they were?

What is a Black Hole ?

JUST WHAT IS A BLACK HOLE?

This was not a simple question. Imagine explaining Schwarzschild's and Oppenheimer's solutions to Einstein's equations, then reviewing how nature squeezes these heavenly bodies until space folds up around them and they disappear . . . all without the use of my hands.
BUT HOW DOES NATURE SQUEEZE THOSE ENORMOUS OBJECTS IN THE SKY?

When a star burns up all its fuel, it collapses due to its own gravity.
YOU KNOW, I'VE ALWAYS WONDERED WHAT STARS WERE MADE OF AND HOW THEY GIVE OFF LIGHT.

The Birth and Death of Stars

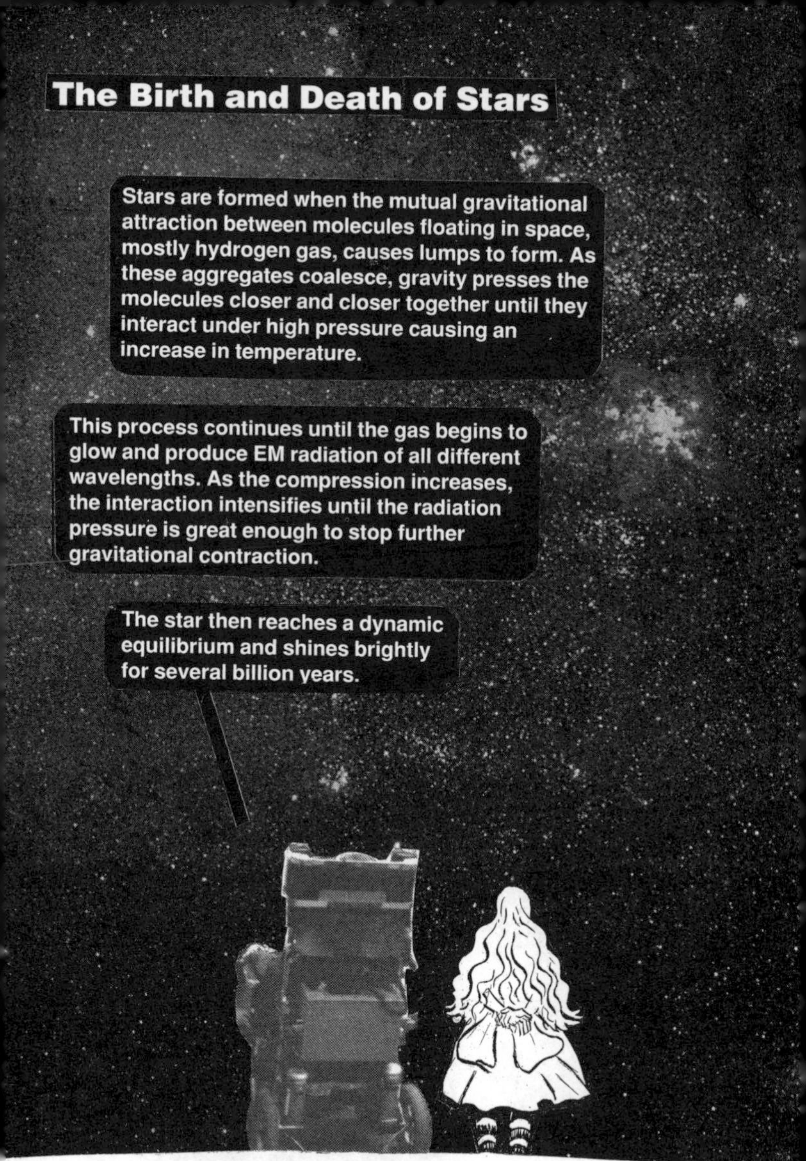

Stars are formed when the mutual gravitational attraction between molecules floating in space, mostly hydrogen gas, causes lumps to form. As these aggregates coalesce, gravity presses the molecules closer and closer together until they interact under high pressure causing an increase in temperature.
This process continues until the gas begins to glow and produce EM radiation of all different wavelengths. As the compression increases, the interaction intensifies until the radiation pressure is great enough to stop further gravitational contraction.
The star then reaches a dynamic equilibrium and shines brightly for several billion years.

BUT WHERE DOES THE STAR GET THE ENERGY TO CONTINUOUSLY HEAT THE MOLECULES AND PRODUCE THE RADIATION?

This problem was first solved by the great English scientist (perhaps the world's first astrophysicist) Sir Arthur Eddington. His famous monograph **On the Internal Constitution of Stars** explained how a star can be fuelled by a reaction at its core providing the energy to heat continually the gas atoms.

HOWEVER, AT THAT TIME I DID NOT UNDERSTAND THE PROCESS AT THE CENTRE OF THE STAR PRODUCING THE RADIATION.
IT WAS LEFT TO THE NUCLEAR PHYSICIST *HANS BETHE* IN 1938.

I SHOWED THAT A NUCLEAR FUSION REACTION (COMBINING HYDROGEN AND DEUTERIUM TO FORM HELIUM) PRODUCES THE ENERGY.
MY MASS/ENERGY EQUATION $E = mc^2$ SHOWS HOW MUCH ENERGY COULD BE PRODUCED AS SOME OF THE STAR'S MASS IS CONVERTED INTO HEAT AND RADIATION, GRADUALLY BURNING ITS FUEL.
AND THEN WHEN THE FUEL IS EXHAUSTED, GRAVITY TAKES OVER AGAIN AND CRUNCHES THE STAR INTO A BLACK HOLE, RIGHT?
Well, it's a little more complicated than that because the final state depends on how massive the star was initially.

How Stars Collapse to Form White Dwarfs, Neutron Stars & Black Holes

Mass of star = **M** in units of solar mass. (If star is 5 times as massive as the Sun, M = 5)

Star burns up all its fuel, hydrogen into helium, and radiation dies out.

Then gravity begins compression again, without resistance.

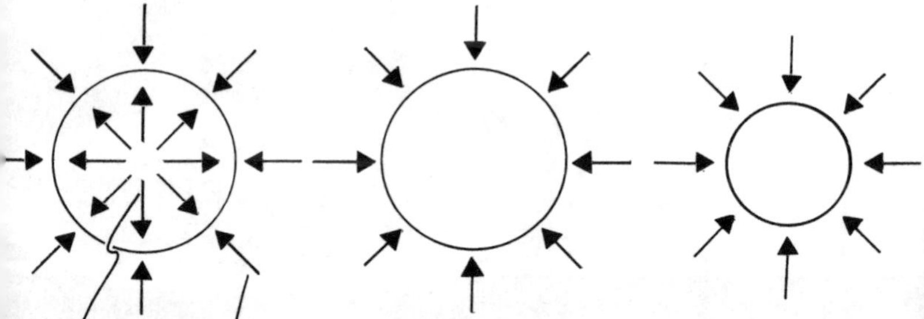

Star burns for billions of years in dynamic equilibrium, giving light and heat.

(Star may explode for a short time to a “red giant” or “supernova”.)

What happens next depends on the initial mass of the star.

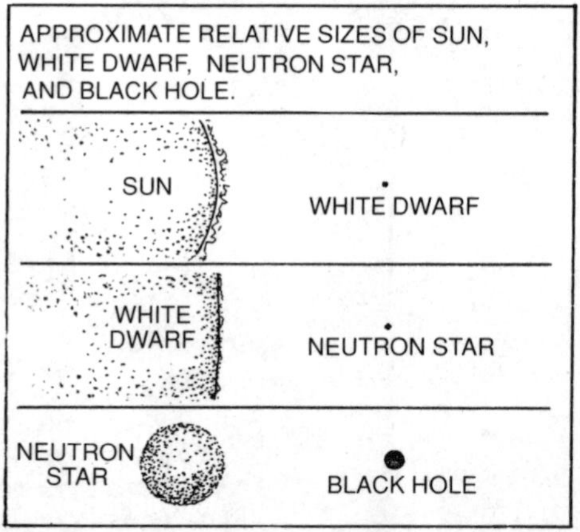

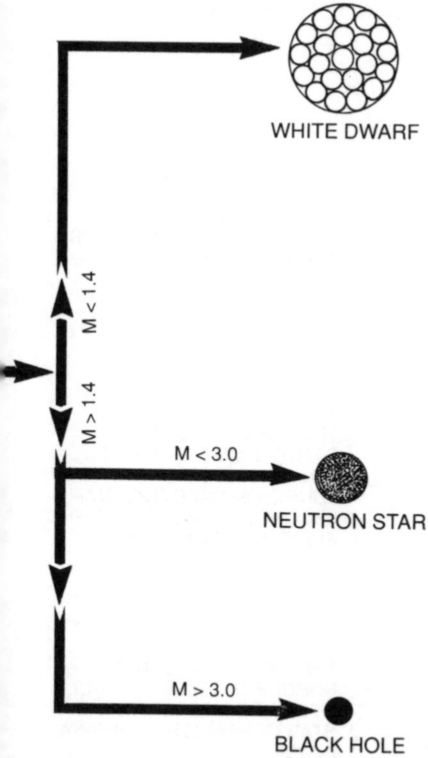

White Dwarf
(radius – 1,600 miles)

If M is less than 1.4, star contracts until gas atoms overlap. Electron repulsion force is enough to stop contraction.

Neutron Star
(radius–16 kilometres)

If M is greater than 1.4, gravity overcomes electrons' heroic stand and pushes them down into the nucleus. The electrons combine with protons to form neutrons. Neutron repulsion stops contraction if M is less than 3.0.

Black Hole
If M is greater than 3.0 (three times solar mass) nothing can stop the contraction. The star collapses completely and disappears from view. A black hole forms.

Traces of some White Dwarfs have been photographed and blips of rotating Neutron Stars can be detected with radio telescopes. But black holes will never be seen directly.

In the black hole case, the space curvature is so extreme that, at a particular radius (called the event horizon), the light from the star's surface is bent in on itself, i.e. the rays actually go *into* the star instead of away from it. The star disappears from view to an outside observer.

These circles of decreasing size show how a very massive burnt-out star, as its diameter decreases, passes through an event horizon to form a black hole, ultimately becoming a singularity at its own centre.

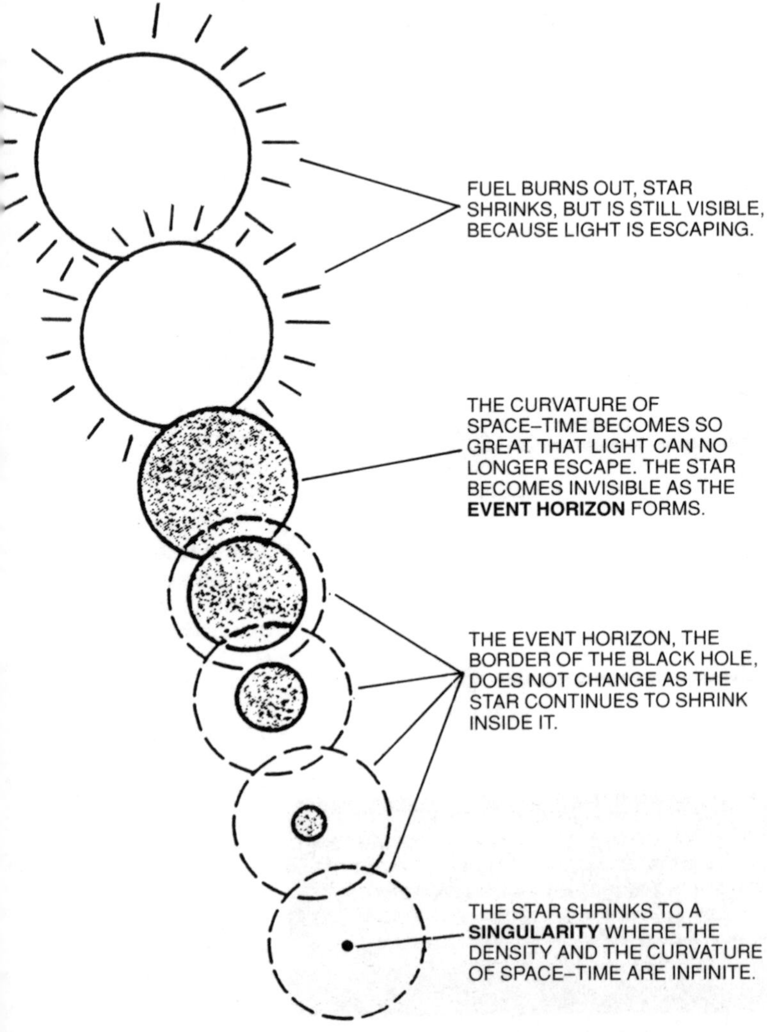

FUEL BURNS OUT, STAR SHRINKS, BUT IS STILL VISIBLE, BECAUSE LIGHT IS ESCAPING.
THE CURVATURE OF SPACE–TIME BECOMES SO GREAT THAT LIGHT CAN NO LONGER ESCAPE. THE STAR BECOMES INVISIBLE AS THE EVENT HORIZON FORMS.
THE EVENT HORIZON, THE BORDER OF THE BLACK HOLE, DOES NOT CHANGE AS THE STAR CONTINUES TO SHRINK INSIDE IT.
THE STAR SHRINKS TO A SINGULARITY WHERE THE DENSITY AND THE CURVATURE OF SPACE–TIME ARE INFINITE.

The illustration below presents the same information in a 3-dimensional diagram which includes **time increasing** in the vertical direction.

This shows the bending of the light paths and indicates how the star's surface has shrunk all the way down to the singularity (right through the event horizon) as the star collapsed.

It is very important to understand the **path** of the light rays from the surface of the star as it passes through the event horizon.

Just before the horizon forms, light rays are bent strongly by gravity and only just leave the star's surface.

A few moments later, when the star is **just inside the event horizon**, the light rays are pulled into the interior of the star towards the singularity at the centre.

But between these two points, when the star **has just reached the event horizon**, gravity is too strong to let light escape but not strong enough to pull the rays into the interior of the star. The light rays hover just at the surface and this defines the event horizon.

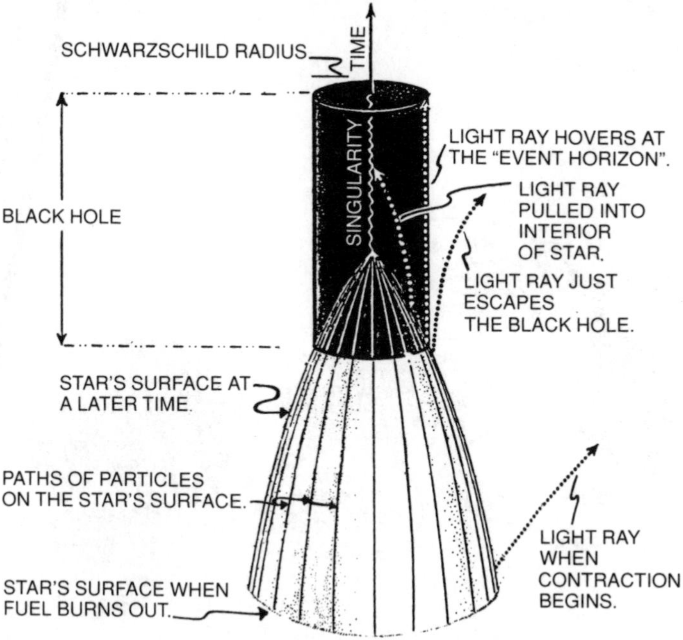

What would happen if someone flew into a black hole?

Einstein and the relativists have an answer that outdoes science fiction. According to the Oppenheimer and Snyder solution, anyone who goes through the event horizon must eventually hit the singularity with disastrous results.

He will be pulled and squeezed – until, at the centre of the black hole, his body would be stretched infinitely long and his width squashed to zero like a length of spaghetti!

Even the atoms in his body would do the same!

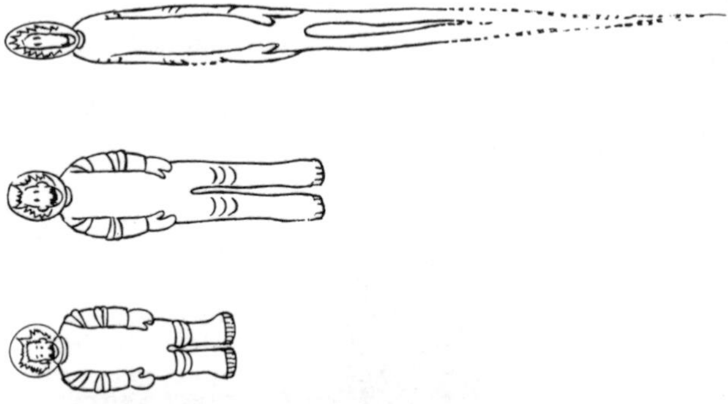

Observational Evidence for Black Holes

Stephen Hawking says there are thousands and thousands of black holes in the Milky Way galaxy alone. But until the day an astronomer is lucky enough to see a well-known star disappear, indirect methods must be used – such as observations on a binary star system with one visible and one invisible (i.e. the black hole) component. John Wheeler has an interesting metaphor for such a system.

In December 1970, the X-ray satellite *Uhuru* was launched from the coast of Kenya. Astronomers were about to use still another part of the EM spectrum – X-rays – to probe the heavens.

Within two years, over 300 sources of X-rays were detected. One of these in the constellation Cygnus (now called Cygnus X-1) looked like just the binary-star system the black hole enthusiasts were waiting for.

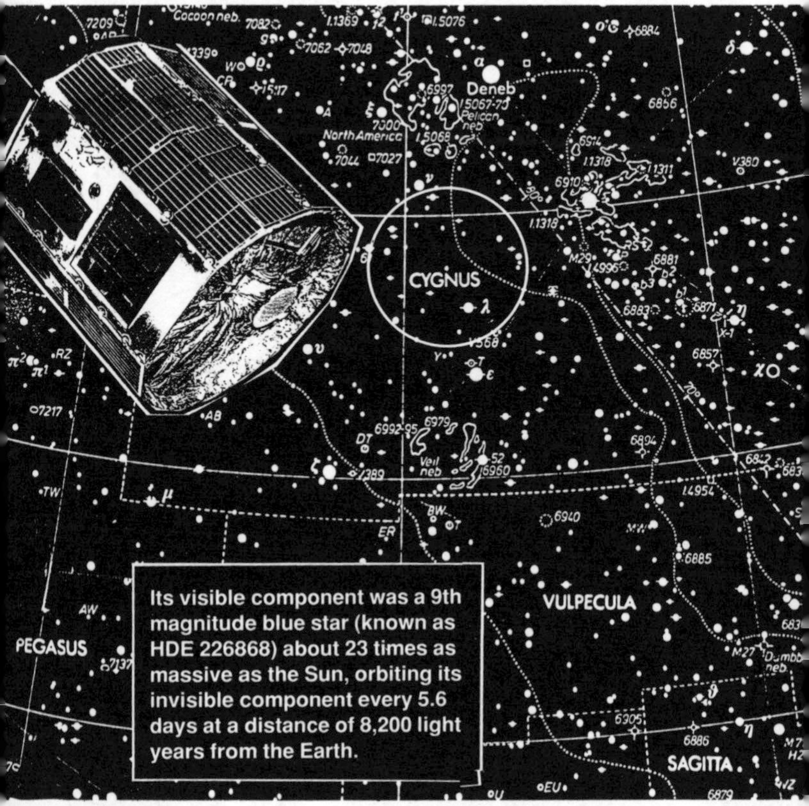
Its visible component was a 9th magnitude blue star (known as HDE 226868) about 23 times as massive as the Sun, orbiting its invisible component every 5.6 days at a distance of 8,200 light years from the Earth.
Cocoon neb.
Deneb
Pelican neb
North America
CYGNUS
X-1
Veil neb
M29
M27
VULPECULA
PEGASUS
SAGITTA

With good estimates of the mass of HDE 226868 and reliable observations of the period of revolution, astronomers could calculate the mass of the invisible component – 10 times as massive as the Sun. **Too big to be a neutron star, it had to be a black hole.**

Theorists quickly developed a model to explain the X-rays. They believe that the black hole is sucking off matter from its visible partner, forming an accretion disk around itself. The hot inner regions, moving close to the speed of light, produce intense bursts of X-rays shortly before the spiralling matter disappears down the hole.

Since the discovery of Cygnus X-1, a second X-ray satellite launched in 1978 called Einstein has mapped out over 1000 X-ray sources. Only two or three are good candidates for black holes, whereas hundreds more have been identified as neutron stars. Nature seems to prefer the stable neutron state to that of the catastrophic black hole.

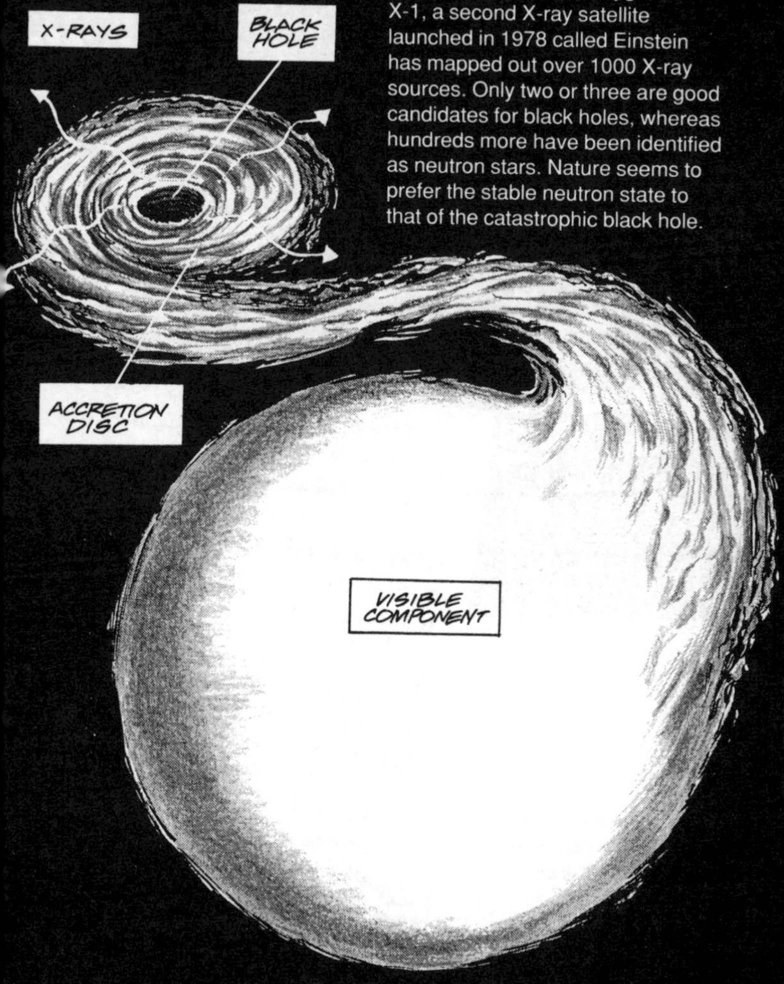

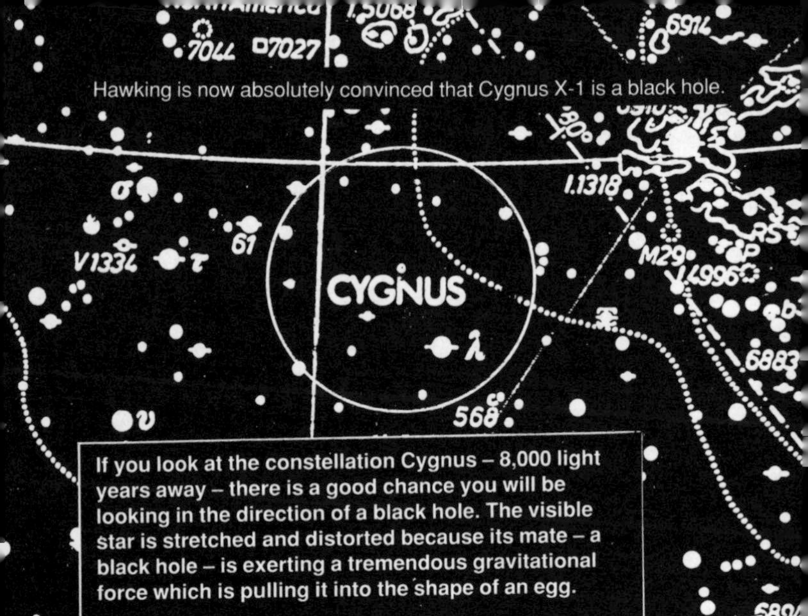

Hawking is now absolutely convinced that Cygnus X-1 is a black hole.

If you look at the constellation Cygnus – 8,000 light years away – there is a good chance you will be looking in the direction of a black hole. The visible star is stretched and distorted because its mate – a black hole – is exerting a tremendous gravitational force which is pulling it into the shape of an egg.

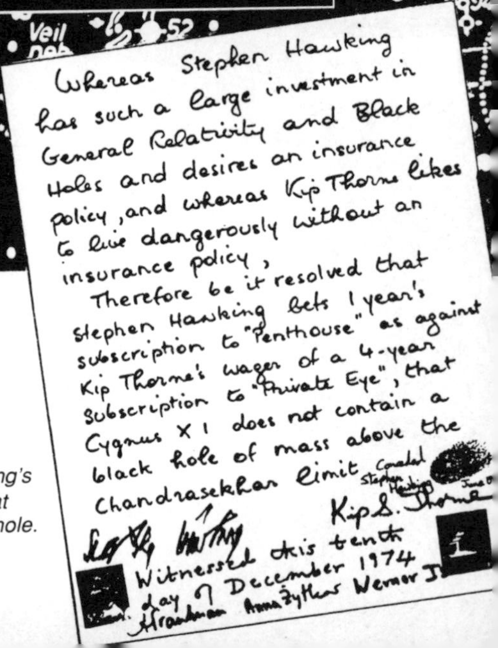

Whereas Stephen Hawking has such a large investment in General Relativity and Black Holes and desires an insurance policy, and whereas Kip Thorne likes to live dangerously without an insurance policy,

Therefore be it resolved that Stephen Hawking bets 1 year's subscription to "Penthouse" as against Kip Thorne's wager of a 4-year subscription to "Private Eye", that Cygnus X1 does not contain a black hole of mass above the Chandrasekhar limit.

Conceded Stephen Hawking

Kip S. Thorne

Witnessed this tenth day of December 1974

Anna Zytkow Werner I

Text of Stephen Hawking's bet with Kip Thorne that Cygnus X-1 is a black hole.

The 1970s: Hawking and Black Holes

By the early 1970s, general relativity and black holes were definitely here to stay. Hawking, by now needing a four-legged walker to get about, was nevertheless poised and ready for action. He was working independently, choosing co-workers from all over the world. He was applying the advanced mathematical techniques introduced by Penrose – mainly from topology – to the properties of black holes. John Wheeler's group at Princeton, Zeldovich and his students in Moscow and Kip Thorne – Wheeler's protégé now at Caltech – could not keep up with him. He managed to master these new methods and stay a step ahead. His name became synonymous with black hole research.

Thorne became a close friend of Stephen and watched his development very closely.

In November 1970, Stephen Hawking was just beginning to reach full stride as a physicist. He had made several important discoveries already, but he was not yet a dominant figure. As the 70s began we watched him become dominant. With his severe disability, how has Hawking been able to out-think and out-intuit his leading colleagues/competitors, people like Roger Penrose, Werner Israel, and Yakov Borisovich Zeldovich?

They had the use of their hands; they could draw pictures and perform many page-long calculations on paper – calculations in which one records many complex intermediate results along the way, and then goes back, picks them up one by one, and combines them to get a final result; calculations that I cannot conceive of anyone doing in his head.

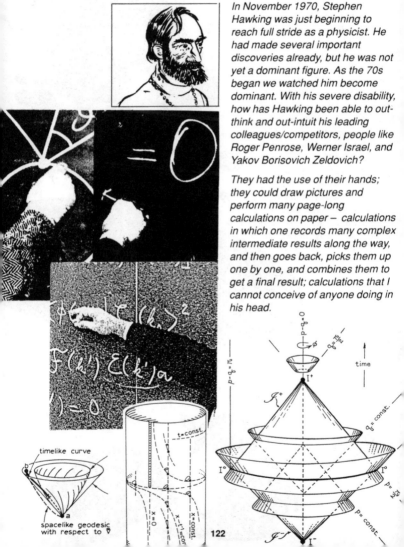

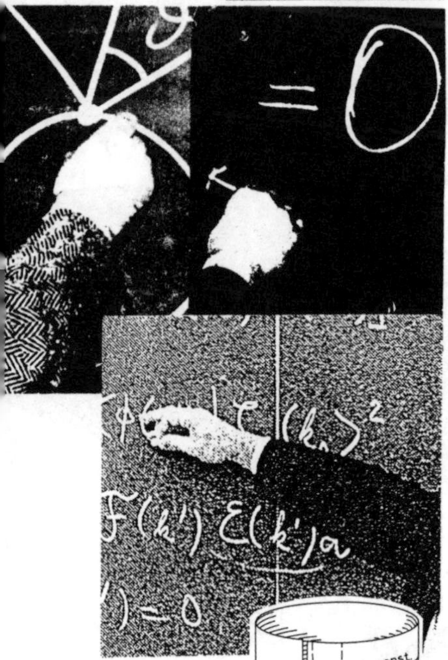

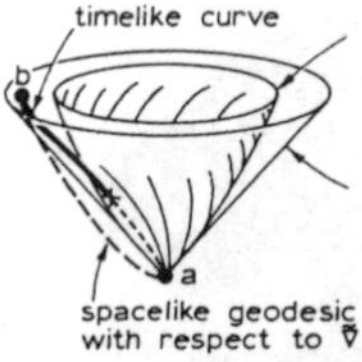

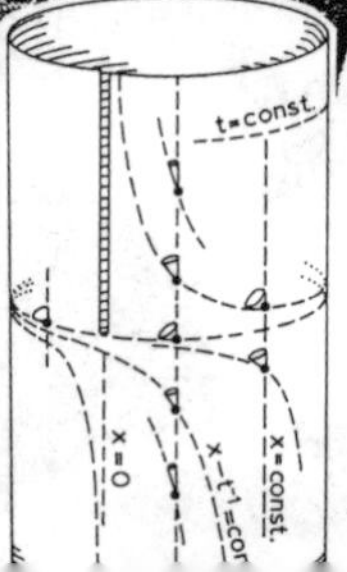

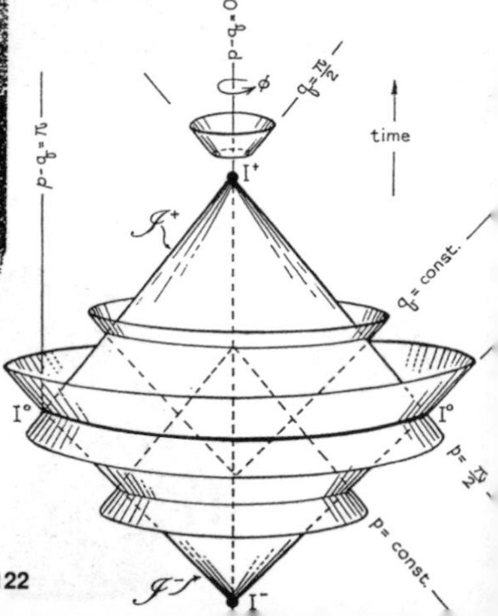

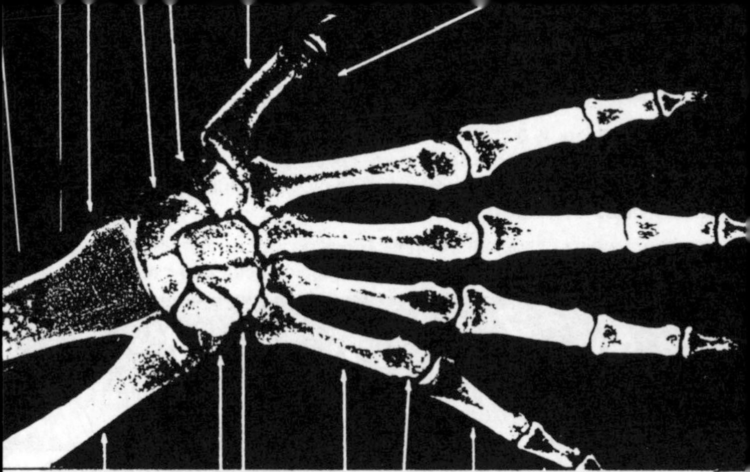

Hawking's mental pictures and mental equations have turned out to be more powerful, for some kinds of problems, than the old paper-and-pens ones, and less powerful for others, and he has gradually learned to concentrate on problems for which his new methods give greater power, a power that nobody else can begin to match.

By the early 1970s, Hawking's hands were largely paralyzed; he could neither draw pictures nor write down equations. His research had to be done entirely in his head. But because the loss of control over his hands was so gradual, Hawking has had plenty of time to adapt. He has gradually trained his mind to think in a manner different from that of the minds of other physicists. He thinks in new types of intuitive mental pictures and mental equations that, for him, have replaced paper-and-pen drawings and written equations.

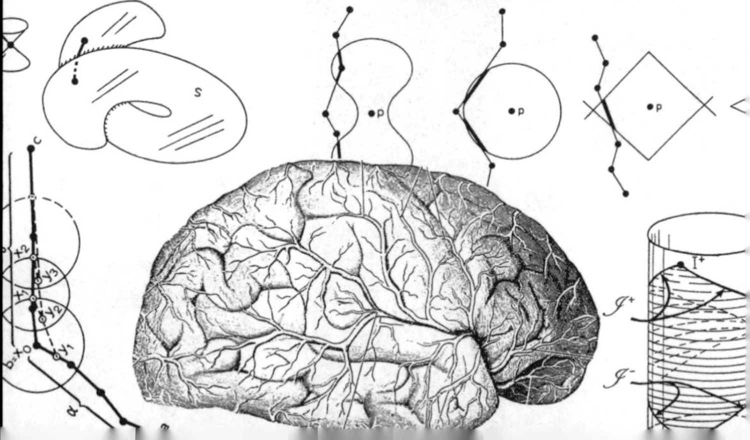

Hawking's Eureka Moment

One of the problems on which Hawking has used mental pictures to gain insight was his study of the surface area of black holes. What started as a rather esoteric problem in black hole dynamics, eventually led to his greatest discovery in physics.

As with Einstein's "happiest thought", Hawking too can remember exactly what he was doing when the germ of his best idea came to him.

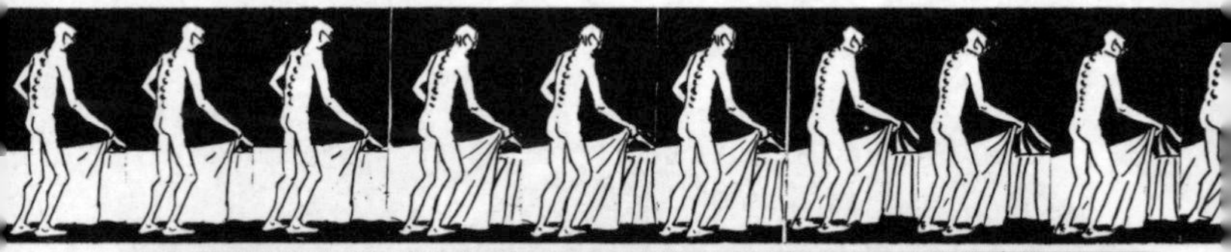

One evening in November 1970, shortly after the birth of my daughter Lucy, I started to think about black holes as I was getting into bed. My disability makes this rather a slow process, so I had plenty of time.

He saw in a flash that **the surface area of a black hole can never decrease**, by considering the paths of light rays hovering just at the event horizon of two black holes. He did not need paper and pen, nor a computer – the pictures were in his head.

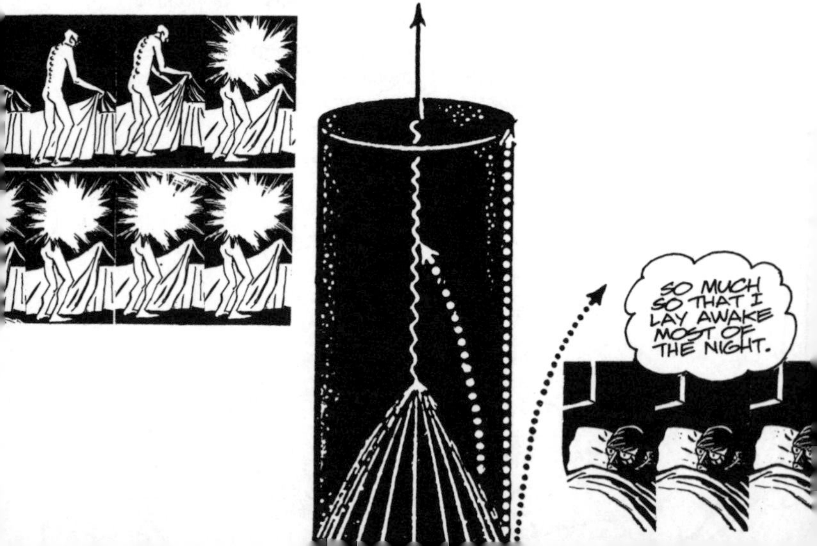

SO MUCH SO THAT I LAY AWAKE MOST OF THE NIGHT.

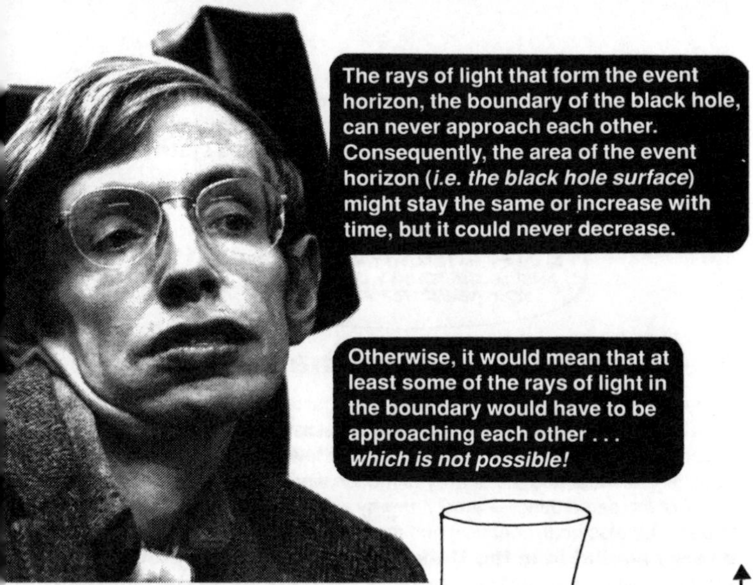

The rays of light that form the event horizon, the boundary of the black hole, can never approach each other. Consequently, the area of the event horizon (*i.e. the black hole surface*) might stay the same or increase with time, but it could never decrease.
Otherwise, it would mean that at least some of the rays of light in the boundary would have to be approaching each other . . . *which is not possible!*

This statement may not seem so remarkable. Since nothing can get out of a black hole and **anything** can go in, how could a black hole get smaller anyway? But Hawking's idea was more general. Even if **two** black holes combine, the total surface area will always be equal to or greater than the sum of the two. It can never decrease. He published his result.

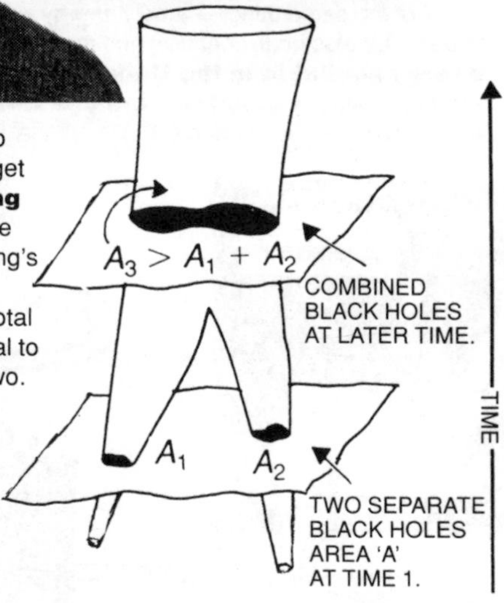

The surface area of a black hole can only stay the same or increase, but can never decrease.

Hawking's Law of Area Increase

Such a statement . . . *can never decrease* . . . immediately gets scientists thinking about the quantity called **entropy** which appears in the second law of thermodynamics: **The entropy (disorder) of a system can only stay the same or increase but never decrease (if the system is isolated and left to reach equilibrium).**

THIS SECOND LAW OF THERMODYNAMICS HAS A VERY INTERESTING HISTORY AND IS CERTAINLY *SOMETHING YOU NEED TO KNOW.*

The Laws of Thermodynamics

During the 19th century, a set of mathematical relationships were developed by chemists, geologists and physicists which combined several seemingly disparate concepts into a few powerful laws. Such quantities as heat and the energy of motion were shown to be different forms of the same thing – namely energy – which had already been used to describe electrical, chemical and magnetic effects. **The total energy available in the Universe (the ultimate isolated system) was a constant and one form could be transformed into another.** This became known as the 1st Law of Thermodynamics.

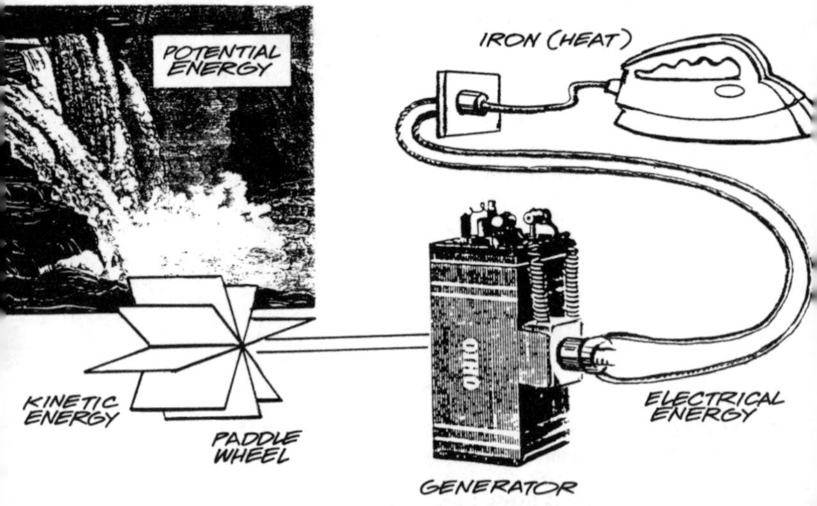
POTENTIAL ENERGY
IRON (HEAT)
KINETIC ENERGY
PADDLE WHEEL
GENERATOR
ELECTRICAL ENERGY

The **2nd Law of Thermodynamics** is more subtle but just as profound. In a lecture delivered in 1854, Hermann von Helmholtz pointed out that as time elapsed all energy would eventually be transformed into heat at a uniform temperature and all natural processes would cease. This is the concept of the *heat death* of the Universe based on the principle of the *dissipation of energy.*

Another way of stating this principle was suggested by the German physicist Rudolf Clausius in 1865.

I INTRO-DUCED A NEW CONCEPT TO WHICH I GAVE THE NAME ENTROPY, DEFINED IN TERMS OF THE HEAT TRANSFERRED FROM ONE BODY TO ANOTHER.

He showed that the total entropy of **a system always increases whenever heat flows from a hot body to a cold body** It also increases whenever mechanical energy is changed into internal (thermal) energy, as in certain collisions and frictional processes.

A more general definition of entropy was proposed by the Austrian physicist Ludwig Boltzmann in 1878.

ACCORDING TO MY DEFINITION, ENTROPY DEPENDS ON THE PROBABILITIES OF MOLECULAR ARRANGEMENTS. FOR EXAMPLE, IF A STATE HAS A VERY LARGE NUMBER OF DIFFERENT WAYS ITS MOLECULES CAN BE ARRANGED, THEN IT HAS A VERY LARGE ENTROPY.
AN EGG FALLS TO THE FLOOR AND BREAKS. IT IS UNLIKELY TO REFORM INTO ITS ORIGINAL SHAPE.

The principle of the dissipation of energy (generalized 2nd law of thermodynamics) can now be stated very simply: *The entropy of an isolated system always tends to increase.* What does this mean?

It means – heat does not flow by itself from cold bodies to hot bodies; a ball cannot bounce higher than its original position by converting heat into mechanical energy; an egg cannot unscramble itself. If the contrary events were to occur, they would not violate any of the principles of Newtonian mechanics – but they would decrease the entropy of a system and are thus forbidden by the 2nd law of thermodynamics. This law tells time which way to go!

How important is this 2nd law of thermodynamics? It should be no less familiar to us than any of the works of Shakespeare, as the writer C.P. Snow remarked in his famous book, **The Two Cultures and the Scientific Revolution.**

Now Back to Black Holes . . .

When bodies reach thermal equilibrium, they **have a temperature** and therefore must emit **thermal radiation**, exchanging energy with their surroundings as described on pages 98 and 99

But everyone knows that black holes do not emit anything – this is the defining characteristic of a black hole. Though anything can fall *into* a black hole, nothing gets out – not even light or any other radiation.

SO IT WAS GENERALLY UNDERSTOOD BY EVERYONE THAT IF BLACK HOLES DON'T RADIATE, THEY CANNOT HAVE A TEMPERATURE, AND THUS CANNOT HAVE ENTROPY. BLACK HOLES ARE CUT OFF FROM THE UNIVERSE AND ARE NOT IN THERMAL EQUILIBRIUM...
OR SO EVERYONE THOUGHT.

That is, until a physics postgraduate student working with John Wheeler in Princeton began to cause trouble.

Controversial Birth of a New Idea

Princeton New Jersey: John Wheeler and postgraduate student Jacob Bekenstein.

Meanwhile back at DAMTP, Stephen Hawking and Brandon Carter are talking about Bekenstein's paper.

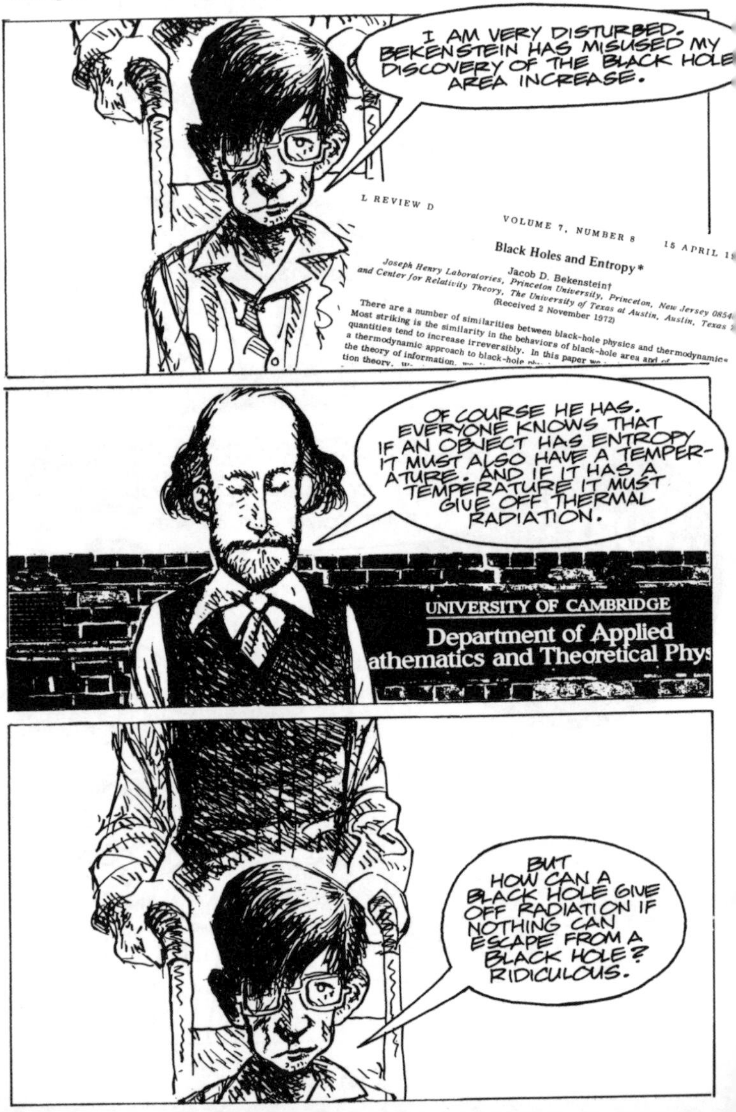

August 1972, Les Houches Summer School on Black Hole Physics

High up on a hillside in the French Alps, Stephen Hawking, James Bardeen and Brandon Carter joined forces to deduce from Einstein's general relativity equations the full set of laws that govern the evolution of black holes. When they were finished, they had produced a set of **laws of black-hole mechanics** that bore an amazing resemblance to the laws of thermodynamics.

S (entropy) = k_1A (surface area of black hole)

T (temperature) = k_2G (surface gravity of black hole)

k_1 and k_2 are constants

EACH BLACK-HOLE LAW, IN FACT, TURNED OUT TO BE IDENTICAL TO A THERMODYNAMIC LAW, IF ONE ONLY REPLACED THE PHRASE "BLACK HOLE SURFACE AREA" BY "ENTROPY," AND THE PHRASE "BLACK HOLE SURFACE GRAVITY" BY TEMPERATURE.
THE COINCIDENCES WERE PILING UP.

Meanwhile, Jacob Bekenstein was a student attending the summer school, still convinced that black holes have entropy.

After the summer school, Bekenstein continued to identify the black hole surface area with entropy in the technical journals. Yet he did not assert that a black hole **has a temperature or that it must emit radiation**. Bekenstein was being inconsistent with the laws of thermodynamics.

Hawking, on the other hand, continued to attack Bekenstein's conclusions, but was becoming increasingly troubled.

SUPPOSE BEKENSTEIN IS RIGHT? I HAVE TO SEARCH FOR A MECHANISM WHICH MIGHT PRODUCE RADIATION FROM A BLACK HOLE.

All the calculations on black holes had been carried out using approximations based on general relativity theory, correct for macroscopic, i.e. large bodies. These approximations ignored any quantum effects, which surely would seem to be negligible in the case of black holes.

THEN HAWKING BEGAN TO EXPLORE THE BOUNDARY BETWEEN THE BLACK HOLE AND THE VACUUM OF INTER-STELLAR SPACE AND WONDERED HOW THE INTENSE GRAVITY AT THE SURFACE MIGHT AFFECT ANY PARTICLES THAT MIGHT APPEAR THERE, WHETHER REAL OR VIRTUAL.
WAIT A SECOND, WHAT IS A *VIRTUAL PARTICLE?*

Time out for **Something You Need to Know**.

The Uncertainty Principle & Virtual Particles

The uncertainty principle, as elucidated by Werner Heisenberg in 1927, states that there are limits on how accurately we can observe certain physical quantities, such as position, momentum, energy and even time. This is not a limit on our measuring instruments but an inherent characteristic of the Universe, which does not reveal **any** quantity with absolute precision.

Think about the vacuum in outer space. We assume it contains absolutely nothing and thus has zero energy. But we can't be sure of this zero energy because of the same argument. Maybe if we look closely enough we can find *some* energy – at least for a short time.

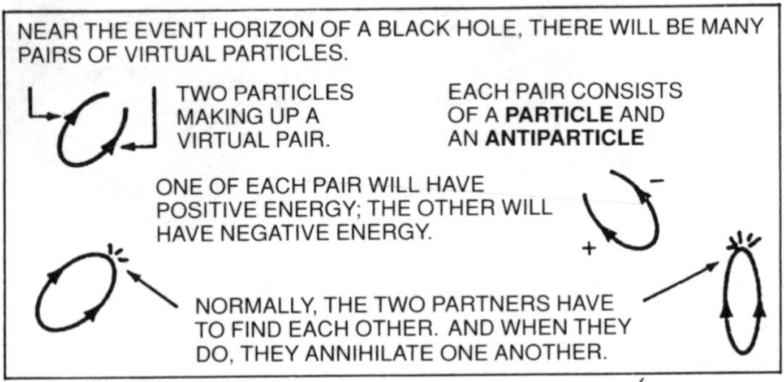
NEAR THE EVENT HORIZON OF A BLACK HOLE, THERE WILL BE MANY PAIRS OF VIRTUAL PARTICLES.
TWO PARTICLES MAKING UP A VIRTUAL PAIR.
EACH PAIR CONSISTS OF A PARTICLE AND AN ANTIPARTICLE
ONE OF EACH PAIR WILL HAVE POSITIVE ENERGY; THE OTHER WILL HAVE NEGATIVE ENERGY.
–
+
NORMALLY, THE TWO PARTNERS HAVE TO FIND EACH OTHER. AND WHEN THEY DO, THEY ANNIHILATE ONE ANOTHER.

THE UNCERTAINTY PRINCIPLE IN FACT PREDICTS THAT ENERGY CAN CONTINUOUSLY APPEAR AND DISAPPEAR ON A SCALE DETERMINED BY PLANCK'S CONSTANT (WHICH IS VERY SMALL). BUT BY EINSTEIN'S EQUATION E = mc², THIS ENERGY CAN TURN INTO PARTICLES AND ANTI-PARTICLES, POPPING IN AND OUT OF EXISTENCE.
THESE ARE CALLED VIRTUAL PARTICLES FLICKERING EVERYWHERE JUST BELOW THE THRESHOLD OF OBSERVABLE REALITY.

Hawking considered what might happen at the surface of a black hole (i.e. at the event horizon), where the intense gravitational field interacts with these virtual pairs. He was in effect combining quantum mechanics and general relativity in a single calculation for the very first time. What he found seemed quite remarkable.

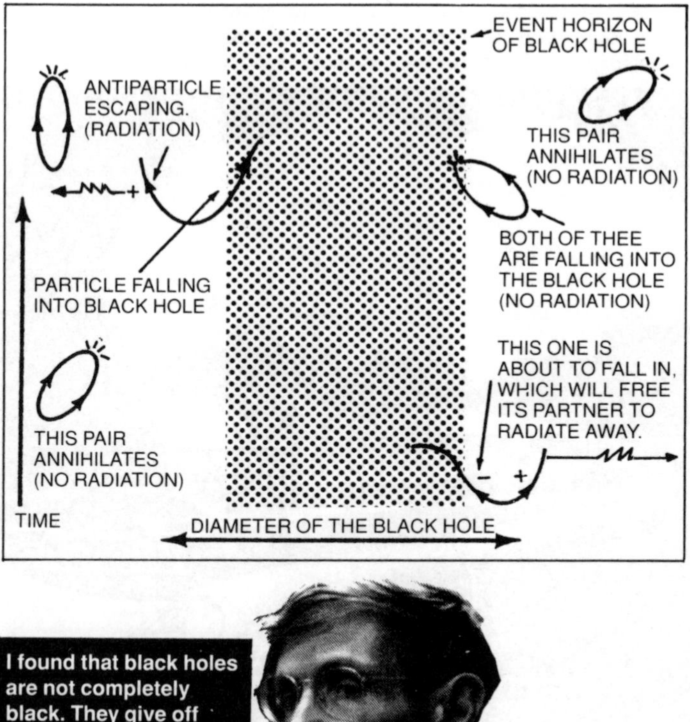

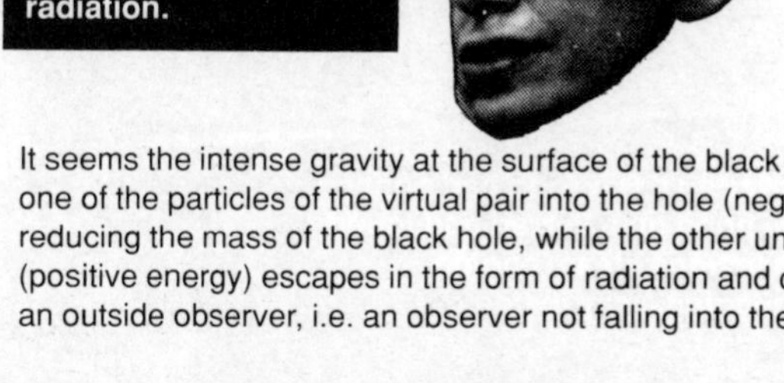

It seems the intense gravity at the surface of the black hole can attract one of the particles of the virtual pair into the hole (negative energy), reducing the mass of the black hole, while the other unpaired particle (positive energy) escapes in the form of radiation and can be detected by an outside observer, i.e. an observer not falling into the black hole.

The most remarkable aspect of this result was the nature of the radiation. It had a perfect *thermal radiation* spectrum which meant that black holes were just like any other body in the Universe. It was now clear that black holes not only have entropy but a temperature as well and obey the classic laws of thermodynamics laid down in the late 19th century.

The science writer Dennis Overbye in his classic book on modern cosmology, **Lonely Hearts of the Cosmos**, produced a wonderful metaphor to describe his feelings about Hawking's discovery.

It was as if Hawking had popped the hood on a Ferrari and found an antique steam engine chugging away inside.

Freeman Dyson, one of the world's top mathematical physicists, was enchanted with the new theory and wrote a popular essay after Hawking visited the Institute of Advanced Study in Princeton.

Hawking was reluctant to publish and had only shared his new results with a few close associates.

Dennis Sciama, visiting Cambridge from Oxford where he had taken an appointment in the physics department, met another of his former students, Martin Rees, then at the Institute of Astronomy in Cambridge.

February 1974, The Rutherford-Appleton Laboratory, Oxford

The chairman, John Taylor, a well-known mathematics professor and writer of a popular book on black holes, introduces Hawking.

Taylor then stormed out of the session. Hawking sat in shocked silence. He knew his talk would be controversial, but he never expected anything like this.

A month after the meeting at Oxford, Hawking published a paper on the new radiation entitled *Black Hole Explosions?* in the journal **Nature**. The paper became the topic of discussion in physics departments everywhere and many were sceptical.

Four months later, Taylor and Paul Davies, a colleague at King's College, London, published a retort in the same journal, *Do Black Holes Really Explode?*

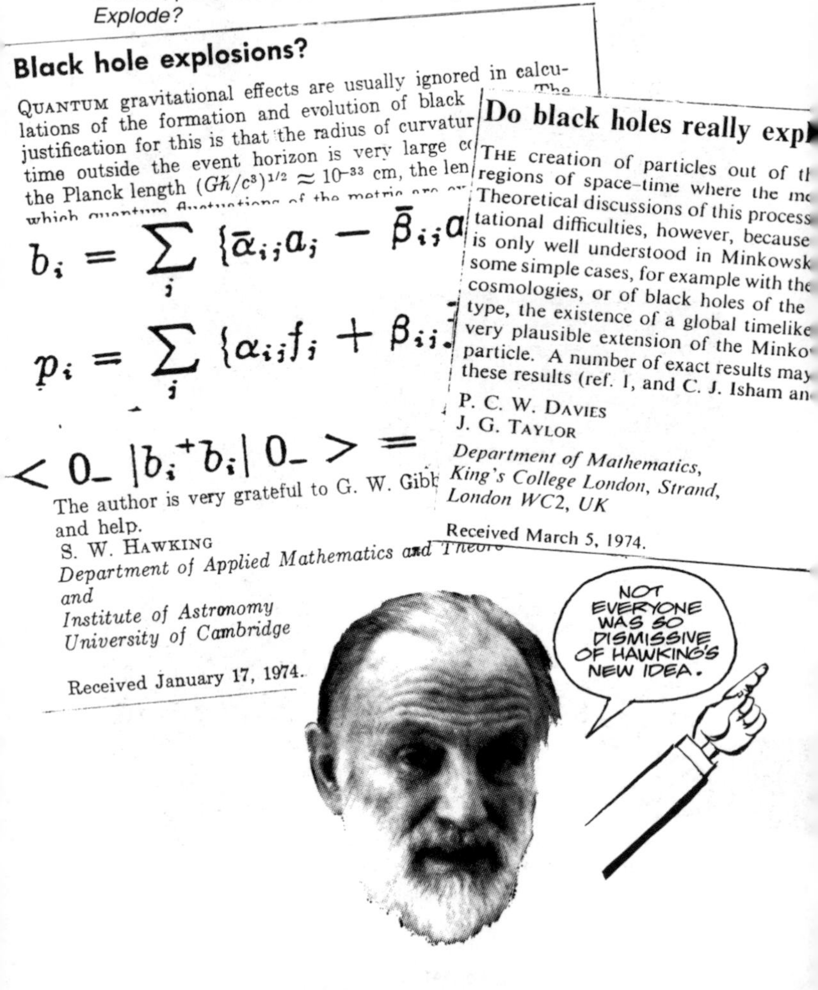

Black hole explosions?

QUANTUM gravitational effects are usually ignored in calculations of the formation and evolution of black … justification for this is that the radius of curvatur… time outside the event horizon is very large c… the Planck length $(G\hbar/c^3)^{1/2} \approx 10^{-33}$ cm, the len…

$$b_i = \sum_j \{\bar{\alpha}_{ij} a_j - \bar{\beta}_{ij} a \ldots$$

$$p_i = \sum_j \{\alpha_{ij} f_j + \beta_{ij} \ldots$$

$$< 0_- | b_i^+ b_i | 0_- > = \ldots$$

The author is very grateful to G. W. Gibb… and help.

S. W. HAWKING
Department of Applied Mathematics and Theor…
and
Institute of Astronomy
University of Cambridge

Received January 17, 1974.

Do black holes really expl…

THE creation of particles out of t… regions of space–time where the m… Theoretical discussions of this process… tational difficulties, however, because… is only well understood in Minkowsk… some simple cases, for example with the… cosmologies, or of black holes of the… type, the existence of a global timelike… very plausible extension of the Minko… particle. A number of exact results may… these results (ref. 1, and C. J. Isham an…

P. C. W. DAVIES
J. G. TAYLOR

Department of Mathematics,
King's College London, Strand,
London WC2, UK

Received March 5, 1974.

Freeman Dyson compares Hawking's formulas to the epoch-making theory of Max Planck in 1900 which led to the quantum theory.

Now Hawking has written down an equation which looks rather like Planck's equation. Hawking's equation is S = kA, where S is the entropy of a black hole, A is the area of its surface, and k is a constant. But what does it really mean to say that entropy and area are the same thing? We are as far away from understanding this now as Planck was from understanding quantum mechanics in 1900. All that we can say for certain is that Hawking's equation is a clue to the riddle of black holes. Somehow, we can be sure, this equation will emerge as a central feature of the still unborn theory which will tie together gravitation and quantum mechanics and thermodynamics.

MAYBE, HAWKING'S NEW THEORY WILL PROVIDE THE CLUE TO QUANTUM GRAVITY.

Perhaps the best way to look at Hawking's discovery is to use another historical analogy. In the year 1900, Max Planck wrote down an equation, E = hv, where E is the energy of a light wave, v is its frequency, and h is a constant which we now call Planck's constant. This equation was the beginning of quantum theory, but in the year 1900 this made no physical sense. It only began to become clear twenty-five years later, when Planck's equation was built into the theory which we now call quantum mechanics.

It is unlikely there has ever been a more powerful demonstration of the self-consistency of physics – a first step towards quantum gravity. It is the unification of three distinct theories of physics which makes Hawking's Radiation so important.

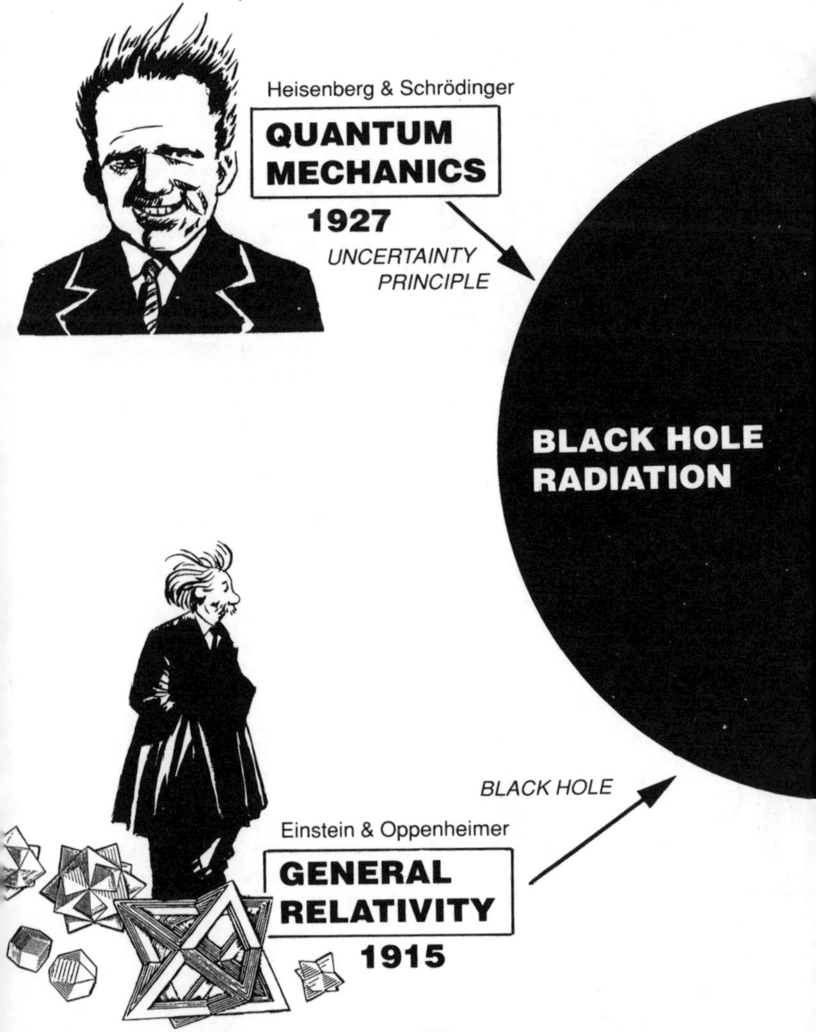

Clausius & Boltzmann
THERMODYNAMICS
2nd LAW OF THERMODYNAMICS (ENTROPY)
(HAWKING 1974)

Recognition of the importance of his work came quickly. Only a few weeks after the paper on black hole radiation was published, Stephen received Britain's highest academic honour. Only 32 years old, he was made a Fellow of The Royal Society, an investiture which made him very proud indeed.

Soon after, Hawking was invited to spend a year away from Cambridge at Caltech, in Pasadena, funded by a special distinguished scholarship, to study cosmology with the eminent American theoretician Kip Thorne.

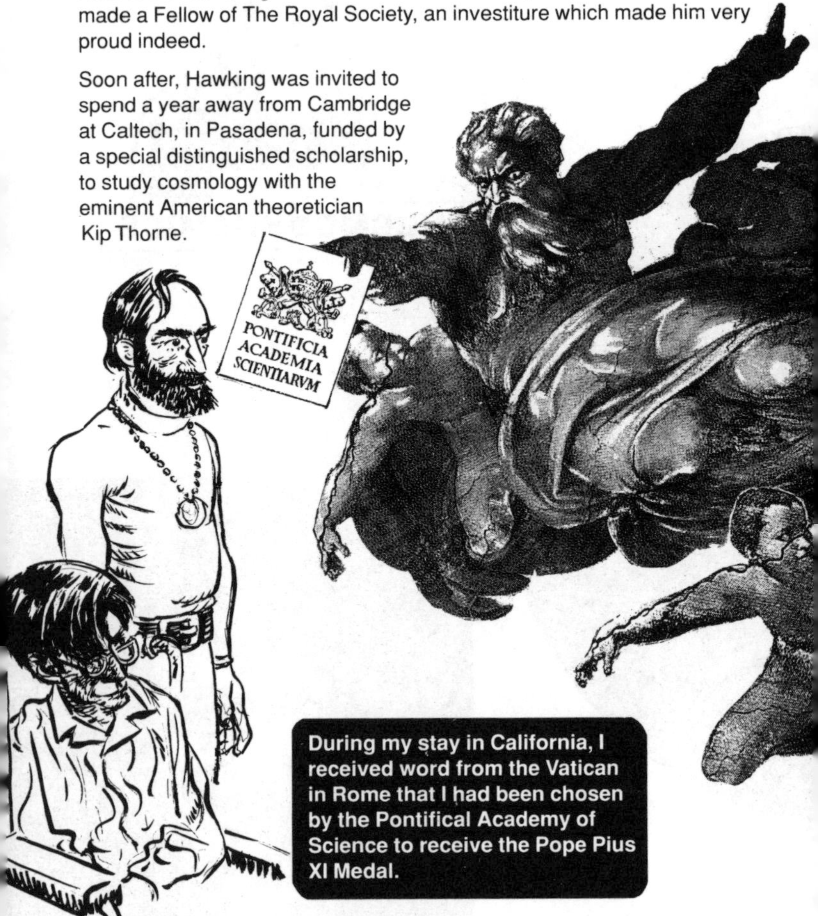

PONTIFICIA ACADEMIA SCIENTIARVM
During my stay in California, I received word from the Vatican in Rome that I had been chosen by the Pontifical Academy of Science to receive the Pope Pius XI Medal.

In a strange way, this award began a shift in his research from black holes to the beginning of the Universe, a subject of great interest to the Roman Catholic Church.

Hawking and the Vatican – a Modern Day Galileo

The powerful Roman Catholic Church has a vested interest in scientific theories about the heavens. For centuries the Church promoted the scientific teachings of Aristotle (a good philosopher but a poor physicist) and the celestial system of Ptolemy which both placed the Earth and Man at the centre of the Universe.

FINALLY IN 1992 THE CHURCH APOLOGISED TO ME. A BIT LATE, *PORCA MISERIA!*

To safeguard the church's teaching, Giordano Bruno was burned at the stake in 1600 for teaching the ideas of Copernicus' **heliocentrism** that the Sun and not the Earth is at the centre of the solar system.

Thirty-three years later, Galileo Galilei was forced to kneel before the Inquisition, with chains of torture rattling in the background, and recant his belief in Copernicanism.
Later, he was placed under house arrest in his villa at Arcetri for the remainder of his days.

The Vatican has since adopted a more subtle approach to scientists who attempt to answer the ultimate questions of the Universe. It now seems happy to court Stephen Hawking, a cosmologist from Protestant England. Why is that?

The Church was quick to accept the idea (i.e. by Vatican standards). On 22 November 1951, at the opening of a meeting of the Pontifical Academy of Sciences, Pope Pius XII declared that Lemaître's idea *accorded with the Catholic concept of creation*. Consequently, any scientist supporting the big bang would certainly be a friend of Rome.

OUR YOUNG FRIEND, IL DOTTORE STEPHEN HAWKING, PROVED IN 1970 THAT EINSTEIN'S GENERAL RELATIVITY DEMANDS THAT ALL THE MATTER AND ENERGY IN THE UNIVERSE MUST AT ONE TIME HAVE BEEN COMBINED IN A *SINGLE POINT* - THE SINGULARITY. PERFETTO!
THAT'S AS CLOSE AS SCIENCE WILL GET TO IDENTIFYING THE HAND OF GOD!
SO IT'S ONLY RIGHT THAT THE PONTIFICAL ACADEMY SHOULD AWARD THE EXCELLENT HAWKING WITH IT'S POPE PIUS XI MEDAL, NO?

Hawking was apprehensive . . .

I was in two minds whether to accept, because of Galileo. When I arrived in Rome to receive the award, I insisted on being shown the record of Galileo's trial in the Vatican Library.

By the late 1970s, Hawking had realized that general relativity is not valid at the moment of the big bang, because of the uncertainty principle, and he was exploring the combination of general relativity and quantum mechanics. He was already beginning to think like a heretic.

But he was back in Rome in 1981, invited to a conference on cosmology organized by the Vatican. By now he had a new area of research, the beginning of the Universe. The paper he gave had a highly technical title.

My interest in the origin and fate of the Universe was reawakened when I attended a conference on cosmology in the Vatican in 1981. Afterwards, we were granted an audience with the Pope, who was just recovering from an attempt on his life.

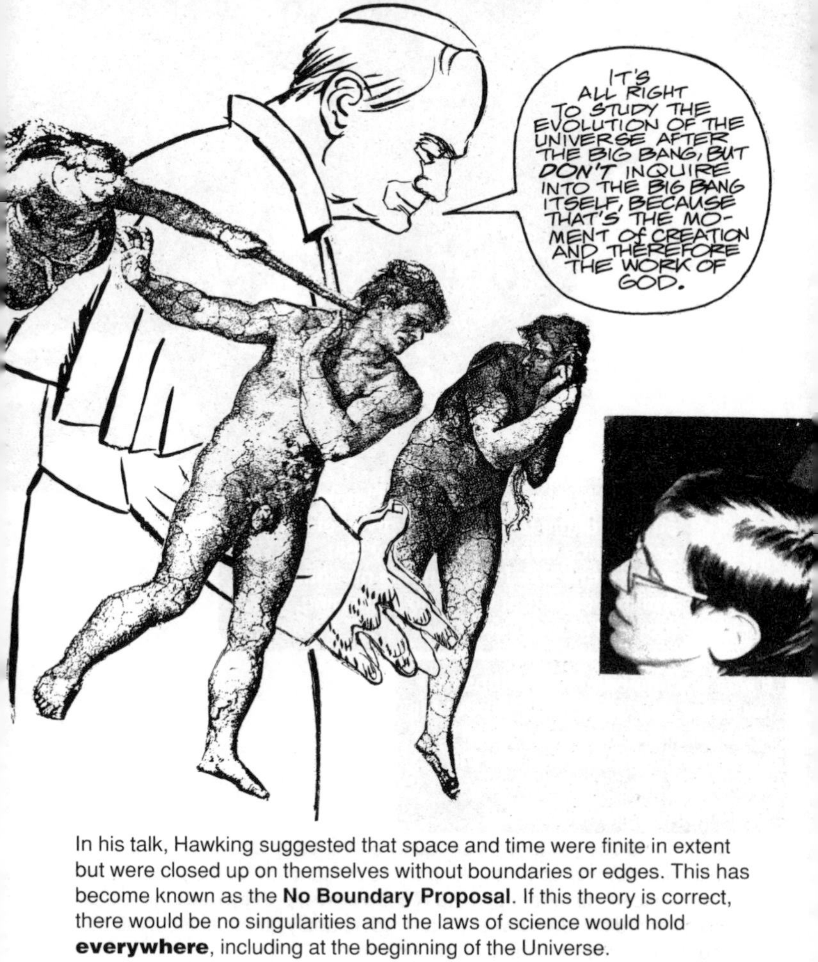

In his talk, Hawking suggested that space and time were finite in extent but were closed up on themselves without boundaries or edges. This has become known as the **No Boundary Proposal**. If this theory is correct, there would be no singularities and the laws of science would hold **everywhere**, including at the beginning of the Universe.

Hawking and the Early Universe

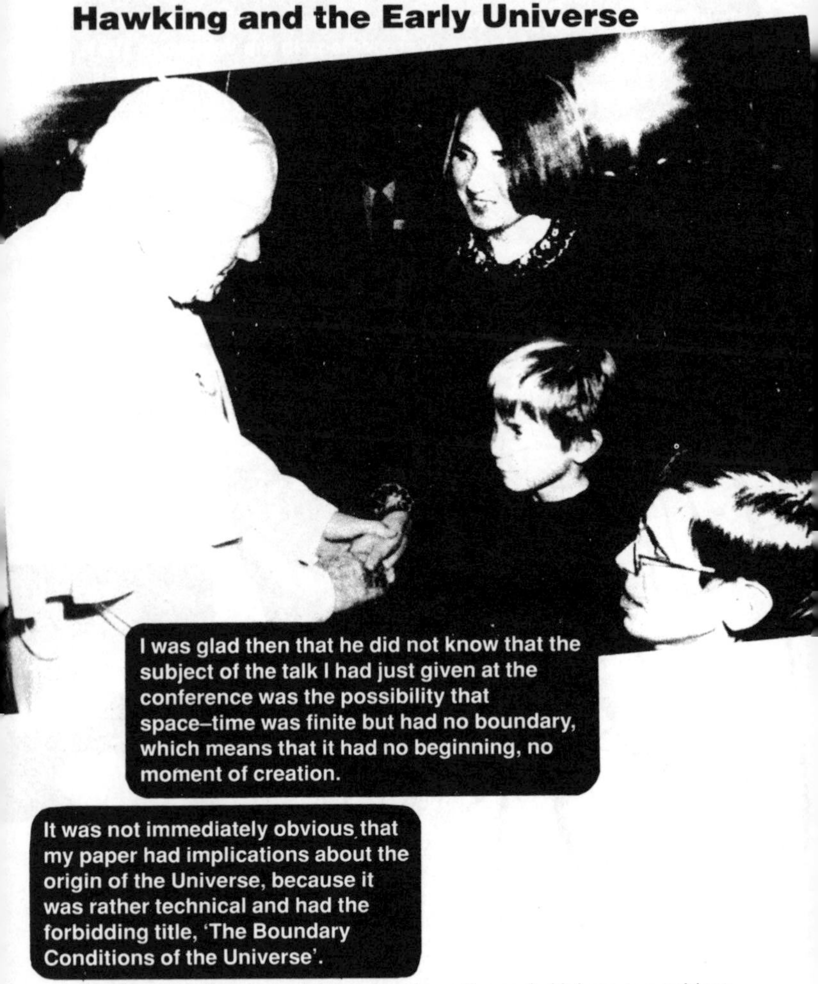

I was glad then that he did not know that the subject of the talk I had just given at the conference was the possibility that space–time was finite but had no boundary, which means that it had no beginning, no moment of creation.

It was not immediately obvious that my paper had implications about the origin of the Universe, because it was rather technical and had the forbidding title, 'The Boundary Conditions of the Universe'.

Hawking had begun to work seriously on the early Universe, a subject which has dominated his thinking to the present day. In the paper he gave at the Vatican, he introduced the **No Boundary Proposal**, his latest and most radical idea. It was an attempt to apply quantum theory to the singularity at the beginning of the Universe.

Why Do We Need Quantum Theory?

In the big bang model of the Universe, the general theory of relativity provides a reliable programme for describing the evolution of our Universe from just moments after time = 0 to the present day. However, thanks to Hawking, we now know that, at the starting point, general relativity predicts a singularity and the theory breaks down. It is a *classical theory* and time and space cannot be described by Einstein's equations when matter is crunched together at such unbelievable densities. **How can physics predict the beginning of the Universe if all the laws break down at the big bang?** Quantum theory **must** be used.

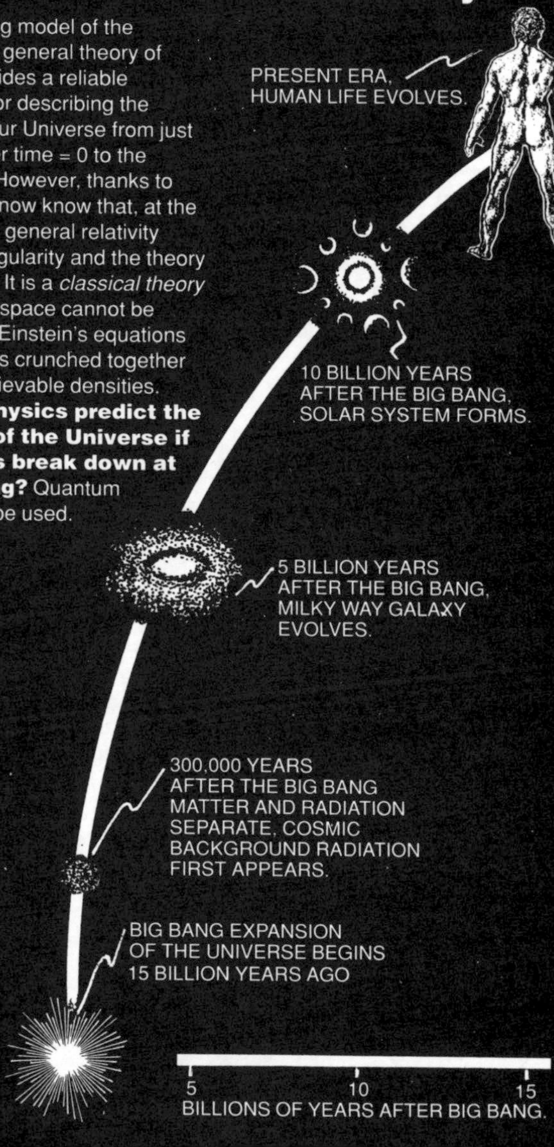

PRESENT ERA, HUMAN LIFE EVOLVES.
10 BILLION YEARS AFTER THE BIG BANG, SOLAR SYSTEM FORMS.
5 BILLION YEARS AFTER THE BIG BANG, MILKY WAY GALAXY EVOLVES.
300,000 YEARS AFTER THE BIG BANG MATTER AND RADIATION SEPARATE, COSMIC BACKGROUND RADIATION FIRST APPEARS.
BIG BANG EXPANSION OF THE UNIVERSE BEGINS 15 BILLION YEARS AGO
5
10
15
BILLIONS OF YEARS AFTER BIG BANG.

Quantum Cosmology

Starting with this question, Hawking and his collaborator, Jim Hartle of the University of California, have used the No Boundary Proposal to develop a new idea in **quantum cosmology**. Unlike previous approaches, Hawking and Hartle (hereafter H & H) have used imaginary time to study the singularity at the big bang.

The reasoning goes like this. At its birth, the Universe is entirely within the quantum state. So H & H treat the Universe as a single quantum system and try to determine its **wave function**. In other words, they are applying standard quantum mechanical principles to the whole Universe "before" the big bang starts.

THIS IS HAWKING'S MOST SERIOUS ATTEMPT TO ACCOMPLISH WHAT EINSTEIN COULD NOT AFTER REACHING THE AGE OF FIFTY, LAY ONE MORE GOLDEN EGG!
ARE YOU LOST? NO WONDER. TRYING TO UNDERSTAND THIS PROPOSAL MAKES THE BIG BANG SEEM LIKE CHILD'S PLAY. BUT LET'S PROCEED...

Quantum Gravity or TOE

The search is called *quantum gravity*, or TOE, the *theory of everything* – a term irritating to most physicists. Attempts so far, by particle physicists and relativists, have yielded few results.

As usual, Hawking is taking a different approach to the problem. Not **quantum gravity**, but his own **quantum cosmology**, finding the wave function for the Universe. This is based on his No Boundary Proposal.

It has always profoundly disturbed me that if the laws of physics could break down at the beginning of the Universe, they could also break down anywhere else. That's why we have developed the No Boundary Proposal which removes the singularity at the beginning of the Universe.

But there is a problem with cosmology because it can not predict anything about the Universe without an assumption about the initial conditions. All one can say is that things are as they are *now* because they were as they were at an earlier stage.

Many people believe that this is how it should be and science should be concerned only with laws which govern how the Universe evolves in time. They feel that the initial conditions for the universe that determine how the Universe *began* is a question for metaphysics or religion rather than science.

YES- LEAVE IT TO RELIGION, AS I SAID BACK IN 1981!

Quantum Cosmology and Complex Time

So what's new about Quantum Cosmology? Well, H & H have used the mathematical trick of complex time to examine **all possible** universes that might form from the initial quantum state. Time is divided into two separate components, one imaginary and one real. Unlike real time, the imaginary component does not vanish at the big bang and the theory is thus useful **at** the singularity. Standard quantum mechanical procedures are then used to arrive at a wave function for the Universe.

COMPLEX TIME NEAR THE BIG BANG SINGULARITY.

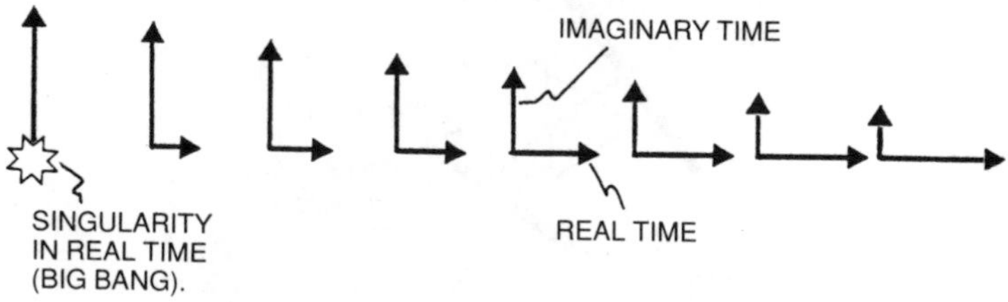

But what are standard quantum mechanical procedures? For that matter, what is a **wave function?**

THIS MOST UBIQUITOUS VARIABLE OF ALL MODERN PHYSICS – THE WAVE FUNCTION – COMES DIRECTLY FROM THE EARLY QUANTUM THEORY.
IT WAS THE BRAINCHILD OF THE VIENNESE PHYSICIST, ERWIN SCHRÖDINGER.

Waves and Particles: Nature's Joke on the Physicists

Experiments have shown that a wave/particle duality exists in Nature. For example, a light beam produces interference effects (acts like a wave) but also kicks electrons out of the surface of a metal (acts like a particle). Similarly, electrons exhibit all sorts of particle properties, yet a beam of electrons produces a diffraction pattern (waves) when sent through a fine comb-like grating. This duality is a basic fact of the physical world and we must live with it. It is a consequence of the well-known *uncertainty principle* . . . or vice-versa.

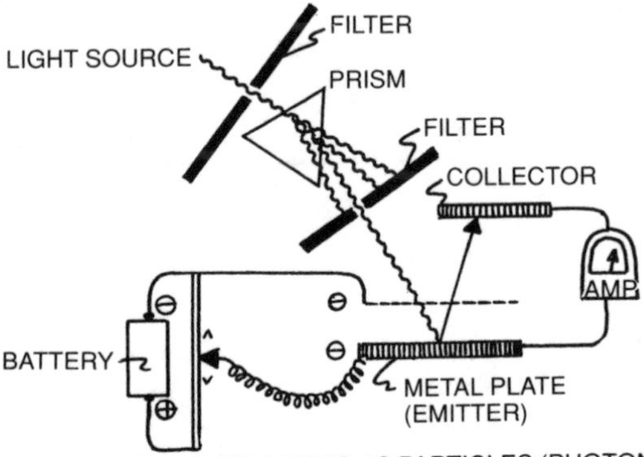

LIGHT WAVES ACTING AS PARTICLES (PHOTONS).

In the 1920s, the early heroes of quantum mechanics – Heisenberg, Schrödinger, Bohr and Born – developed a mathematical language which described both properties – wave and particle – at the same time. The most elegant form of this language was an equation due to Schrödinger, the solution of which – the wave function – determines the behaviour of a system of **particles**.

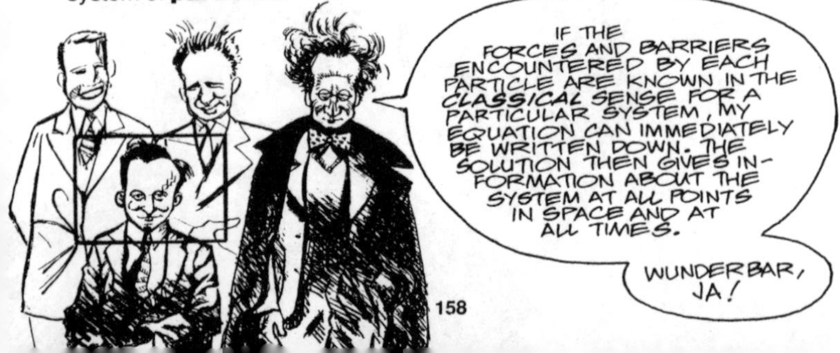

IF THE FORCES AND BARRIERS ENCOUNTERED BY EACH PARTICLE ARE KNOWN IN THE CLASSICAL SENSE FOR A PARTICULAR SYSTEM, MY EQUATION CAN IMMEDIATELY BE WRITTEN DOWN. THE SOLUTION THEN GIVES INFORMATION ABOUT THE SYSTEM AT ALL POINTS IN SPACE AND AT ALL TIMES.
WUNDERBAR, JA!

The Strange World of Quantum Mechanics

But what is a wave function? What exactly is waving?

Here is what Max Born proposed (ironically, following an idea of Einstein's).

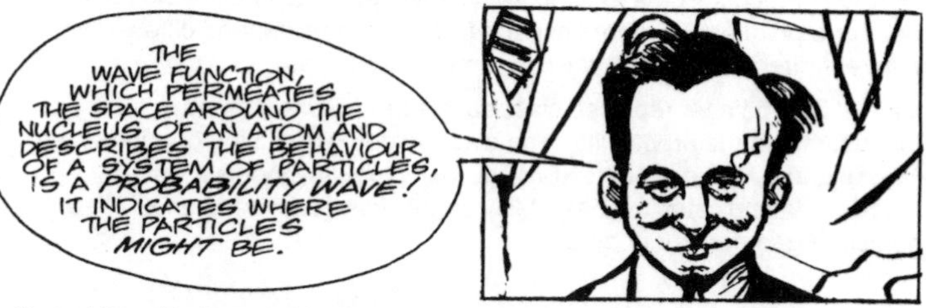
THE WAVE FUNCTION, WHICH PERMEATES THE SPACE AROUND THE NUCLEUS OF AN ATOM AND DESCRIBES THE BEHAVIOUR OF A SYSTEM OF PARTICLES, IS A *PROBABILITY WAVE!* IT INDICATES WHERE THE PARTICLES *MIGHT* BE.

One of the simplest problems to solve using quantum mechanics is the hydrogen atom. When the Schrödinger equation is solved for this case, the resulting wave function determines the probability of each energy state of the atom since it gives the probability of finding the electron at various distances from the nucleus. The nucleus is enveloped in a probability cloud, instead of precise planetary-type electron orbits of the classical atom.

CLASSICAL PICTURE OF HYDROGEN ATOM.

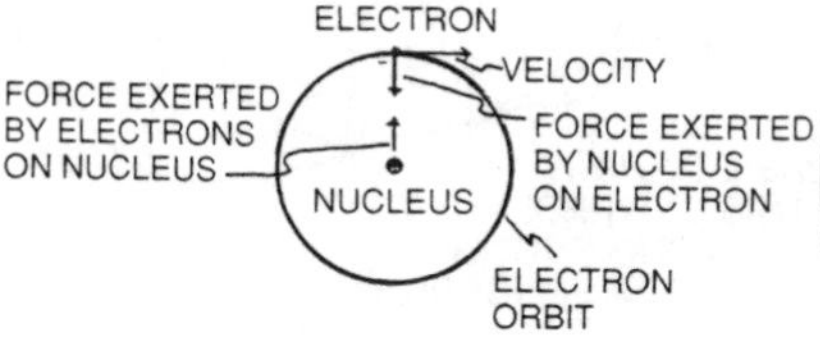
ELECTRON
VELOCITY
FORCE EXERTED BY ELECTRONS ON NUCLEUS
FORCE EXERTED BY NUCLEUS ON ELECTRON
NUCLEUS
ELECTRON ORBIT

WHERE THE PROBABILITY CLOUD SURROUNDING THE HYDROGEN ATOMIC NUCLEUS IS DENSE, ONE IS MORE LIKELY TO FIND THE ELECTRON, BUT ONE CAN NEVER SAY EXACTLY WHERE IN THE ATOM THE ELECTRON IS LOCATED AT ANY ONE INSTANT. ALL ONE CAN SPECIFY IS THE PROBABILITY THAT IT WILL BE IN VARIOUS PLACES.

QUANTUM PICTURE OF HYDROGEN ATOM.

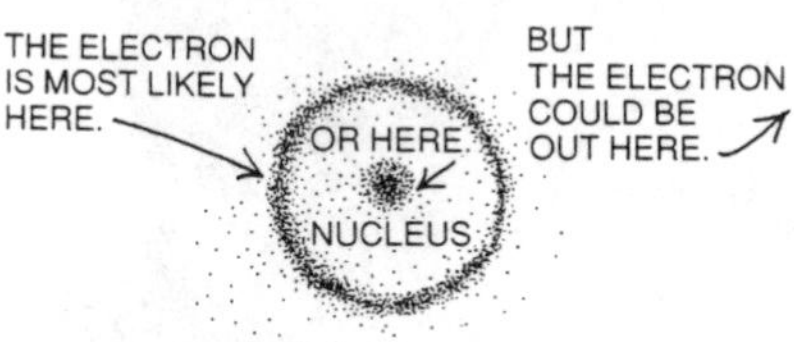
THE ELECTRON IS MOST LIKELY HERE.
OR HERE.
NUCLEUS
BUT THE ELECTRON COULD BE OUT HERE.

Quantum Cosmology: Applying Schrödinger's Equation to the Universe

Is Hawking a bold thinker? Instead of **electron orbits** in the atom, think of **cosmological models** of the Universe. General relativity allows a variety of models: some expand from a point to a maximum size, then back to a point again; others expand forever; others expand differently in different directions. Yet all satisfy Einstein's equations.

Just as Schrödinger replaced classical electron orbits with wave functions that described the probability of an electron doing one thing or another, so Hawking and Hartle assign individual cosmological models a wave function that indicates the probability of the Universe having one particular geometry or another.

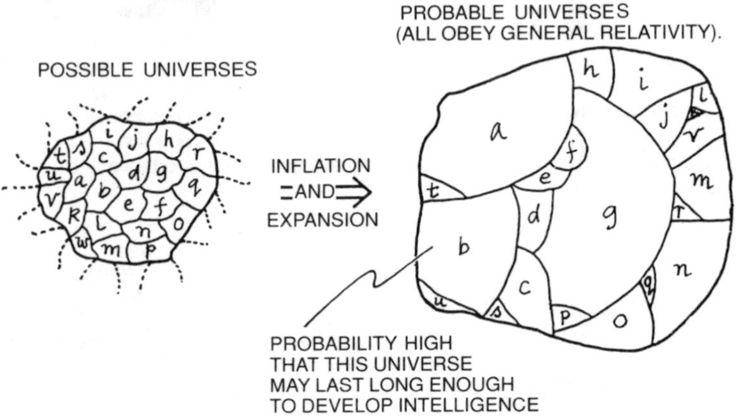
PROBABLE UNIVERSES
(ALL OBEY GENERAL RELATIVITY).
POSSIBLE UNIVERSES
INFLATION
AND
EXPANSION
PROBABILITY HIGH
THAT THIS UNIVERSE
MAY LAST LONG ENOUGH
TO DEVELOP INTELLIGENCE

BY CHOOSING ONLY UNIVERSES WITH NO BOUNDARIES – EITHER IN SPACE OR TIME – H & H OBTAIN RESULTS WHICH SEEM CONSISTENT WITH OBSERVATIONS ON OUR UNIVERSE.

Closed universes satisfy this restriction. They are finite but have no edges, something like the two-dimensional surface of the Earth. They expand, come to a halt, then fall back to the same state like the points on the rim of the bowl shown in the sketch.

Depicted in this way, **closed** universes would have a beginning and an end, and would therefore have boundaries only in **real** time. The imaginary component, however, is continuous. So, H & H make the initial and final singularities of the closed Universe disappear.

They also demonstrate that **uniform** universes are the most probable and end up predicting that our Universe is both **closed** and **uniform** – a finite sphere of space–time with no edges.

Map of uniform distribution of galaxies – from NASA INFRARED SATELLITE.

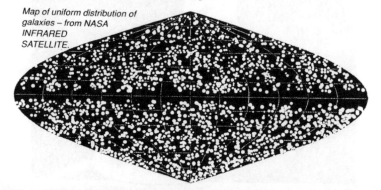

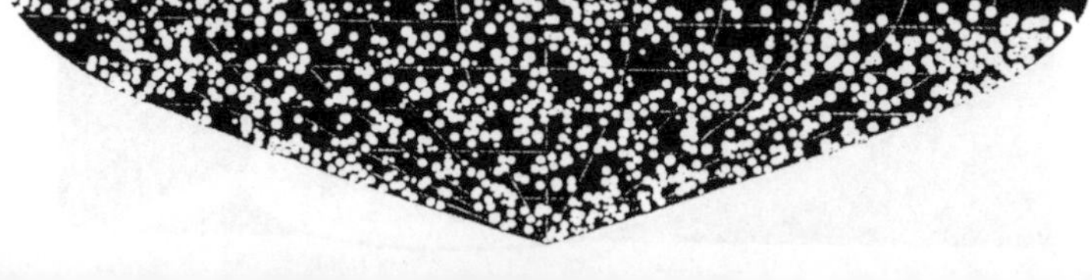

DAMTP: 17 February 1995

As Hawking told the author only six weeks before this book was published . . .

The No Boundary Proposal predicts a Universe that starts out in a very smooth and ordered way. It expands by *inflation* first, then goes over to the standard *hot big bang* model, further expanding to a maximum radius before collapsing to a *big crunch* singularity in a disordered and irregular way.

Calculations carried out so far on simple models indicate that a No Boundary Proposal Universe would be very much like our own. In addition, it would incorporate certain important ideas from contemporary cosmology – such as **inflation** and **quantum fluctuations**. Even the **anthropic principle** seems to fit. If you can understand these last three concepts, you should have a very good picture of Stephen Hawking's Universe. Not bad for a beginner!

Inflation

In the late 1970s, a new concept of **inflation** was introduced which proposed that the Universe expanded from an initial state smaller than a proton to a macroscopic size about ten metres across in only a fraction of a second. The rate of expansion was enormous. The idea solved two problems which had been nagging cosmologists for years.

1. Why is the Universe so flat, i.e. shows no evidence of curvature?
2. Why is the cosmic background radiation so uniform?

1. The first of these questions implies that the mass density of the Universe is perfectly tuned to the critical value from its earliest expansion, a mind-boggling proposition (see page 49). But a rapid expansion at the beginning would flatten out the Universe to the critical mass density as a simple diagram can show.

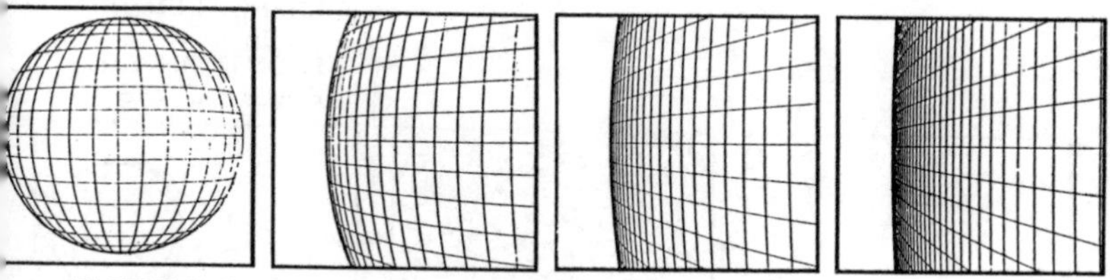

FLATTENING OF THE UNIVERSE BY INFLATION

2. Inflation can also explain why the background radiation is so uniform. When the Universe was of infinitesimal size, all matter and energy was homogeneous, since everything was connected to everything else. As inflation took place, the homogeneity that existed at that early instant was spread across the much larger Universe, which continued to expand. Thus, when matter and radiation de-coupled about 300,000 years later, the Universe was still amazingly uniform.

Inflation and Quantum Fluctuations

The inflation that smoothed out the early Universe could also produce small density variations which might explain galaxy formation. Recall from our discussion of virtual particles on page 136 that if we look closely enough at any physical system – even a vacuum – we observe the effects of **quantum fluctuations.**

Inflation does not **erase** these quantum fluctuations but establishes them as **density variations** which appear as ripples of matter–energy across space–time. These ripples should then be imprinted on the background radiation as tiny temperature variations.

These temperature variations are precisely what George Smoot and his Berkeley–NASA team were looking for with the Cosmic Background Explorer Satellite (COBE) experiment launched in 1989. We need one more bold concept . . .

THE FIRST FRACTION OF A SECOND.

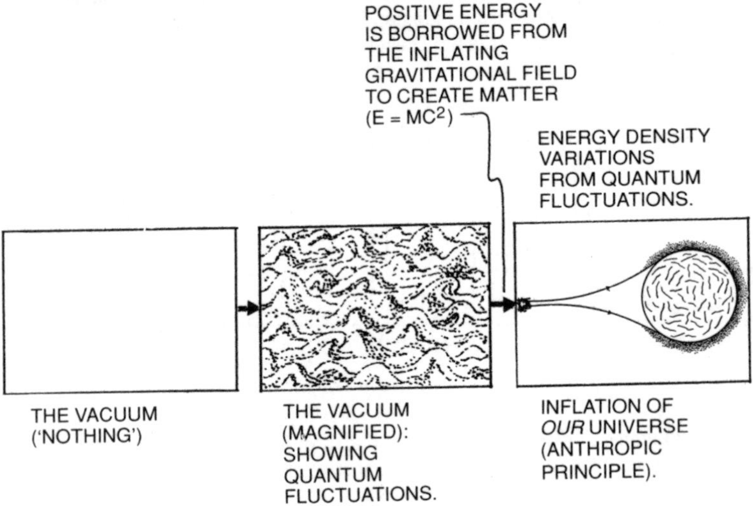

The Anthropic Principle

The anthropic principle is a quasi-metaphysical notion which implies that, if a particular universe does not take on fundamental constants of Nature which allow for the existence of life and the development of intelligence, there will never be anyone to report on its properties. That is why our Universe seems so right to us, it's tuned perfectly.

Although many scientists rubbish this idea, no less an authority than Nobel Laureate Steven Weinberg (who wrote the seminal book on the early Universe, **The First Three Minutes**) believes that **quantum cosmology** provides a context in which the anthropic principle becomes simple common sense. **The most probable universe is the one that we're in!** As Voltaire's absurd philosopher Pangloss keeps telling Candide, "We live in the best of all possible worlds."

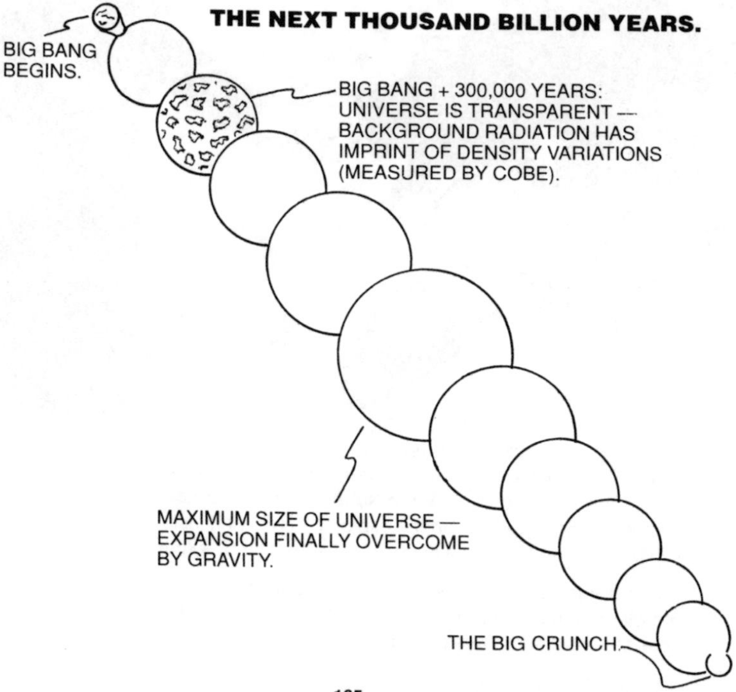
THE NEXT THOUSAND BILLION YEARS.
BIG BANG BEGINS.
BIG BANG + 300,000 YEARS: UNIVERSE IS TRANSPARENT — BACKGROUND RADIATION HAS IMPRINT OF DENSITY VARIATIONS (MEASURED BY COBE).
MAXIMUM SIZE OF UNIVERSE — EXPANSION FINALLY OVERCOME BY GRAVITY.
THE BIG CRUNCH.

Hawking's Nobel Prize

Stephen Hawking has received just about every award and honour which can be given to a scientist. Naturally, the question arises whether he will be awarded the most famous of all – an invitation to the Royal Academy of Sciences in Stockholm to receive the Nobel Prize in Physics.

There are complications. First of all, the award is only rarely given for work in astronomy or cosmology rather than pure physics. The second obstacle is more serious. Alfred Nobel was a very practical man (he made his fortune from patents on the explosive TNT) and insisted that to be eligible, theoretical discoveries must be verified by experiment.

For cosmologists like Hawking, whose laboratory extends to the most remote regions of the Universe, experimental verification may never be possible or, at best, take decades.

Let's review Hawking's major theoretical discoveries which might win him the Nobel Prize.

1. Using General Relativity, Hawking and Penrose showed that the classical concept of time must have begun with a singularity at the Big Bang and thus the Universe existed at one time in a hot, dense state.

2. In 1974, he discovered that black holes radiate like thermodynamic bodies (now called **Hawking Radiation**) and possess a temperature (proportional to their surface gravity) and an entropy (proportional to their surface area).

3. He presented a model for the early Universe called the **No Boundary Proposal** with Jim Hartle which predicts density variations in the early Universe due to quantum fluctuations of the vacuum.

Ironically, **Hawking Radiation**, his most significant work, seems an unlikely candidate for the Nobel award as it seems impossible to detect.

However, both the Big Bang singularity (hot, dense state of the Universe) and quantum fluctuations (seeds for galaxy formation) could be proved if very accurate **absolute** and extremely sensitive **differential** measurements were made of the cosmic background radiation.

That is exactly what the COBE project did between 1989 and 1992.

COBE: the Greatest Discovery of All Time (?)

COBE took twelve years to design and carry out, but the results were nothing short of spectacular. Launched in 1989, the instruments took only 8 minutes to verify the conclusions based on the 1964 measurements of Penzias and Wilson, but this time at many different wavelengths. The data traced out a near perfect thermal radiation curve (see page 99) for a background temperature of 2.736 degrees C above absolute zero.

This was COBE I which used an *absolute* microwave radiometer calibrated by a bath of liquid helium on board the satellite. The results proved without a doubt that the detectors were looking at the remnant of the hot, dense state of the early Universe which we call the big bang. Such a curve would have thrilled Max Planck, as it did the American Astronomical Society when first presented in 1990.

COBE MEASUREMENTS OF BACKGROUND RADIATION.

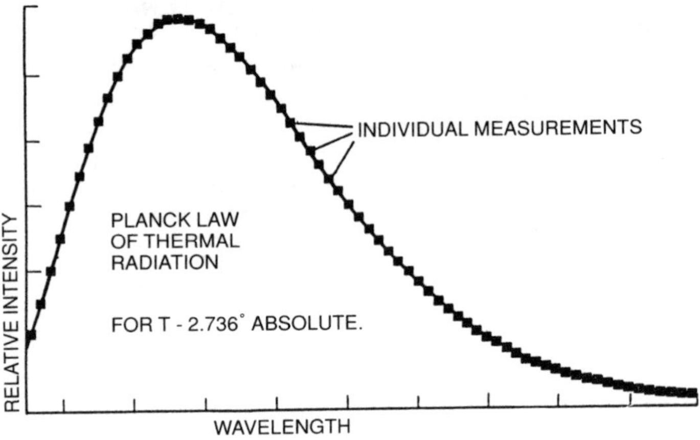

But the big news was still to come. COBE II used a sensitive *differential microwave radiometer* (DMR) which doesn't measure the absolute temperature of the radiation at a given point in the sky; rather, it measures the **difference** in temperature between two points. The COBE I single antenna gives the answer: "The temperature at point A is 2.725 degrees." But the COBE II dual-antenna *differential* radiometer gives the answer: "The temperature difference between point A and point B is 0.002 degrees."

THE COBE SPACECRAFT

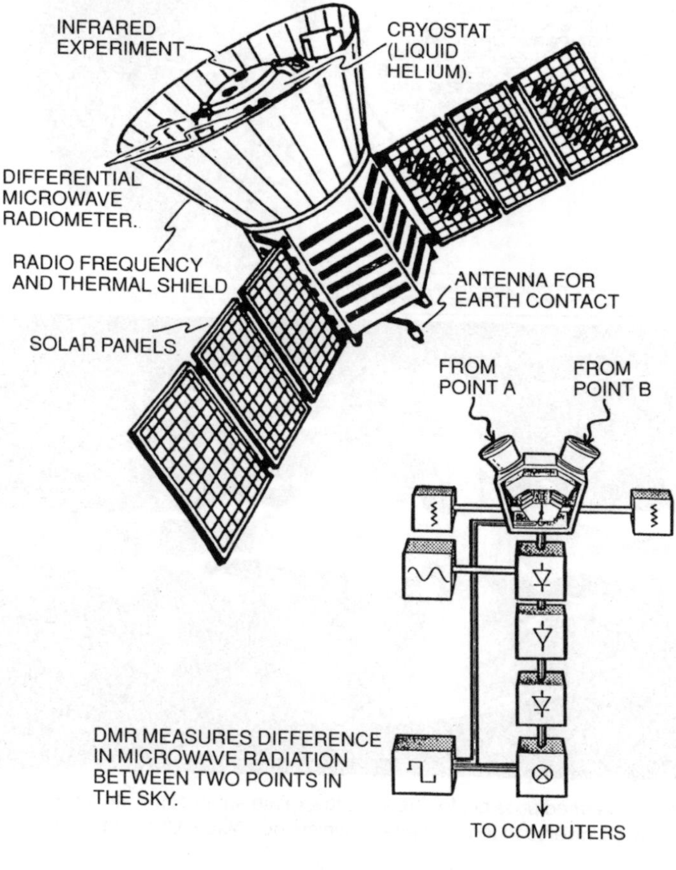

This was George Smoot's project – to look for evidence of ripples in the space–time of the 300,000-year-old Universe. In April 1992, after more than two years of data collecting and analysis, Smoot and his team made a dramatic announcement. The COBE satellite had detected tiny temperature variations of the order of about one-hundred-thousandth of a degree in the background radiation.

ACCORDING TO COMPUTER GENERATED PLOTS OF THE ENTIRE SKY, THE TEMPERATURE WAS MINUTELY HIGHER IN THE DIRECTION OF THE LARGE GALACTIC CLUSTERS AND SLIGHTLY LOWER IN THE GREAT COSMIC VOIDS.

COBE MAP OF THE MICROWAVE SKY SHOWING OUR GALAXY AND COSMIC RIPPLES.

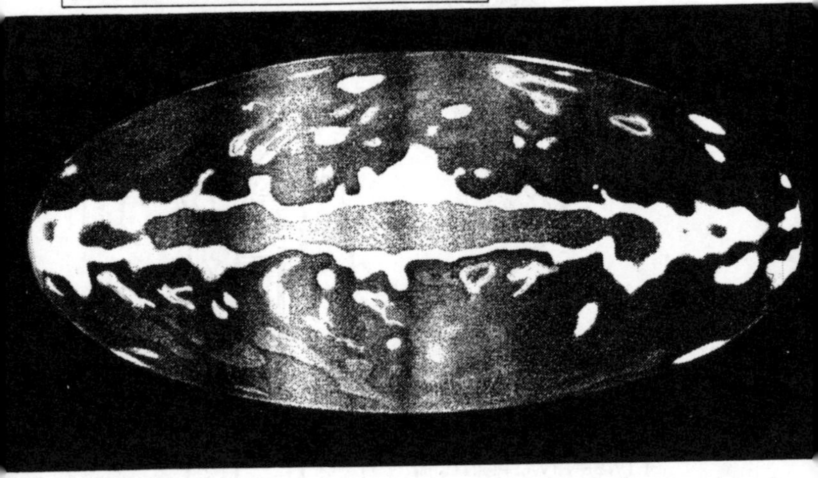

It now seemed possible for theorists to explain some of the structures seen in today's Universe in terms of events which took place billions of years ago.

The report was greeted with an enthusiastic media response all over the world.

NASA

The Cosmic Background Explorer mapped the sky and measured the Big Bang's background radiation.

"It's the greatest discovery of the century – if not of all time"
IF YOU'RE RELIGIOUS, IT'S LIKE SEEING GOD.

Both Hawking and Smoot made statements which together just about covered the two ends of the emotional spectrum. Smoot is a religious man and has accepted the big bang as a creation event. COBE's results moved him emotionally.

Hawking sees things differently. To him, the variations in the background radiation seen by COBE are simply evidence for the presence of quantum fluctuations in an inflationary Universe consistent with his No Boundary Proposal. Any wonder he's smiling.

COBE's success is seen by most scientists to be a stunning confirmation of big bang cosmology. But the game is not yet up. The final solution to the mysteries of the beginning and structure of the Universe may be much more complicated.

The Earth-centred cosmos of Aristotle and Ptolemy, the Sun-centred system of Copernicus, Le Maître's Cosmic Egg and Hawking's No Boundary Proposal are just steps along the way to deeper understanding of the Universe and our place in it. The journey is everyone's to contemplate, to understand, to enjoy.

Stephen William Hawking, cosmologist — an example of *homo sapiens circa* AD 2000 — certainly has done his part.

Further Reading

Books about Hawking

Stephen Hawking, A Life in Science, Michael White and John Gribbin, New York, Plume Books (Putnam).

Stephen Hawking, Quest for a Theory of Everything, Kitty Ferguson, New York, Bantam Books 1992.

Development of Classical Astronomy

The Sleepwalkers, Arthur Koestler, New York, Grosset & Dunlap 1959.

Coming of Age in the Milky Way, Timothy Ferris, New York, Anchor Books 1989.

New Cosmology/Black Holes

A Brief History of Time, Stephen Hawking, New York, Bantam Books 1988.

The First Three Minutes, Steven Weinberg, New York, Bantam Books 1984.

Black Holes and Time Warps, K. S. Thorne, New York, W. W. Norton & Co. 1994.

Black Holes and Warped Spacetime, W. M. Kaufmann, San Francisco, W. H. Freeman 1979.

Black Holes, Jean Pierre Luminet, New York, Cambridge University Press 1992.

In Search of the Big Bang, John Gribbin, New York, Bantam Books 1986.

Lonely Hearts of the Cosmos, Dennis Overbye, New York, Harper Collins 1991.

COBE

Wrinkles in Time, George Smoot and Keay Davidson, New York, Avon Books 1994.

Acknowledgements

I was encouraged at the outset by a conversation with Dennis Sciama last summer in Oxford and by the publication of Kip Thorne's book which was heaven sent. John Gribbin's books and his regular column in the *New Scientist* have been very helpful. Special thanks to John Wheeler, Freeman Dyson, Jacob Bekenstein and Simon Shaffer for useful points of reference and graphics. I discussed quantum cosmology with Chris Isham and he also read a draft manuscript just before going to press.

My time at DAMTP was essential to develop a fresh insight into the Hawking story. The staff in the department, Ms Sue Masey in particular, were most cooperative with my many requests. Jane Hawking provided some helpful background material and also read the manuscript before publication.

On the home front my family has been most supportive, especially my wife Pat, who for the past seven months has given me the psychological and intellectual space I needed to tell this story. She knows about such things.

My colleague Maryke Brecher covered up for my mysterious disappearances when the going got tough and Kathy Black helped me meet (almost) every deadline with her formidable keyboard skills.

A very special thanks to Stephen Hawking who endorsed this project from the beginning and slotted me into his tight schedule on numerous occasions for discussion and advice. Stephen affects most people who work with him and I am no exception. Watching him communicate has taught me to be more concise, more accurate and more clear in everything I do. Also, I hope I shall never allow trivial ailments like a headache or fatigue to interrupt any worthwhile endeavour.

Oscar Zarate has made this book different from any other book on physics or astronomy that I have ever seen. The goal was to illustrate every idea I deemed important, no matter how abstract – to get both sides of the brain working. To achieve this, he read dozens of books on physics and astronomy and listened with rapt interest to my late night discourses as the book's outline took shape. I hope it has worked!

J.P. McEvoy, London, March 1995

Oscar Zarate thanks Judy Groves for her help with the diagrams, Woodrow Phoenix for his lettering, Marta Rodrigues for screening the photos and Bill Mayblin for his advice on the graphics.

J.P. McEvoy received his Ph.D. in physics at Imperial College, University of London, 1968. For twenty-five years, active in physics research and teaching at RCA, Clark University and the American School in London, he has published over 50 technical papers and has recently been involved in science journalism and multi-media development for educational television. He is also the author of *Introducing Quantum Theory.*

Oscar Zarate has illustrated introductory guides to Freud, Quantum Theory, Mind & Brain, Machiavelli, Melanie Klein, Lenin and the Mafia. He has also produced many acclaimed graphic novels, including *A Small Killing*, which won the Will Eisner Prize for best graphic novel of 1994, and has edited *It's Dark in London*, a collection of graphic stories, published in 1996.

Photography by Mark McEvoy and David Simmonds.

Typesetting by Wayzgoose.

Index